Practical Guide to Disruption and Productivity Loss on Construction and Engineering Projects

Practical Guide to Disruption and Productivity Loss on Construction and Engineering Projects

Roger Gibson

WILEY Blackwell

This edition first published 2015
© 2015 by John Wiley & Sons, Ltd.

Registered Office
John Wiley & Sons, Ltd, The Atrium, Southern Gate, Chichester, West Sussex,
PO19 8SQ, United Kingdom.

Editorial Offices
9600 Garsington Road, Oxford, OX4 2DQ, United Kingdom.
The Atrium, Southern Gate, Chichester, West Sussex, PO19 8SQ, United Kingdom.

For details of our global editorial offices, for customer services and for information about
how to apply for permission to reuse the copyright material in this book please see our website at
www.wiley.com/wiley-blackwell.

The right of the author to be identified as the author of this work has been asserted in accordance with the
UK Copyright, Designs and Patents Act 1988.

All rights reserved. No part of this publication may be reproduced, stored in a retrieval system, or
transmitted, in any form or by any means, electronic, mechanical, photocopying, recording or otherwise,
except as permitted by the UK Copyright, Designs and Patents Act 1988, without the prior permission of
the publisher.

Designations used by companies to distinguish their products are often claimed as trademarks. All brand
names and product names used in this book are trade names, service marks, trademarks or registered
trademarks of their respective owners. The publisher is not associated with any product or vendor
mentioned in this book.

Limit of Liability/Disclaimer of Warranty: While the publisher and author(s) have used their best efforts
in preparing this book, they make no representations or warranties with respect to the accuracy or
completeness of the contents of this book and specifically disclaim any implied warranties of
merchantability or fitness for a particular purpose. It is sold on the understanding that the publisher is not
engaged in rendering professional services and neither the publisher nor the author shall be liable for
damages arising herefrom. If professional advice or other expert assistance is required, the services of a
competent professional should be sought.

Library of Congress Cataloging-in-Publication Data

Gibson, Roger, 1944– author.
 Practical guide to disruption and productivity loss on construction and engineering
projects / Roger Gibson.
     pages   cm
   Includes index.
   ISBN 978-0-470-65743-0 (hardback)
 1. Construction contracts–Great Britain.   I. Title.
   KD1641.G53 2015
   343.4107′8624–dc23
                                                            2014032112

A catalogue record for this book is available from the British Library.

Wiley also publishes its books in a variety of electronic formats. Some content that appears in print may
not be available in electronic books.

Set in 11/13pt Minion by SPi Publisher Services, Pondicherry, India
Printed and bound in Malaysia by Vivar Printing Sdn Bhd

1   2015

Contents

Preface		vii
Acknowledgements		ix
1	**Introduction**	**1**
	1.1 Introduction	1
	1.2 The aims of this book	3
	1.3 The SCL Protocol	3
	1.4 Conclusion	9
2	**Contracts and Case Law**	**11**
	2.1 Introduction	11
	2.2 Contracts	13
	2.3 Case law	18
3	**Planning, Programmes and Record Keeping**	**61**
	3.1 Background and history of planning	61
	3.2 Planning and programming	64
	3.3 Programme submission, review and acceptance	78
	3.4 Programme updates and revisions	84
	3.5 Progress records and other record keeping	91
4	**Delay, Disruption and Causation**	**99**
	4.1 Delay	99
	4.2 Disruption	107
	4.3 Causation	118
5	**Loss of Productivity**	**123**
	5.1 Introduction	123
	5.2 Productivity and efficiency	124
	5.3 Common causes of loss of efficiency	128
	5.4 Methods of productivity measurement	133

6 Acceleration and Mitigation   **153**
    6.1   Acceleration   153
    6.2   Mitigation   164

Appendix 1   Definitions and Glossary   169
Appendix 2   Levels of Programmes   177
Appendix 3   SCL Protocol; Guidance Clauses on 'Disruption'   187
Index   191

Preface

I have been involved in the construction industry in the UK and overseas for over 50 years, both at project level in planning and project positions and in head-office organisations in managerial roles. During this time, and in particular during the last 20 years which I have spent primarily in time-related disputes and claims, I have become increasingly aware of the lack of a comprehensive, easy to understand, practical and 'down to earth' reference book for those involved in the preparation and assessment of disruption and loss of productivity claims.

The views expressed by me in this book represent many years' experience of looking at projects that have gone wrong and resulted in a dispute(s) between the parties. In practice, many projects are completed without major claims, and where these do occur they are settled promptly and professionally without escalating into a formal dispute. Unfortunately, a claim that evolves into a formal dispute often stretches the resources of the parties and their consultants and can add financial pressure in resolving the dispute.

Many construction firms, large and small alike, lack staff with the skills required to produce well-presented disrupted and loss of productivity claims. Similarly, the receiving party, architect, engineer or employer, often does not have the in-house skills to review such submissions and claims thoroughly, and delays making a proper decision or resorts to external consultants for assistance.

Roger Gibson
Spring 2014

Acknowledgements

I am indebted to my family and Dayna for their encouragement and support during the writing of this book. Thanks are also due to my past and present colleagues who have offered numerous helpful suggestions.

Finally, the views expressed in this work are my own and I take full responsibility for them.

Chapter 1
Introduction

1.1 Introduction

This book is a practical text that seeks to demystify the measurement of site labour/resource productivity.

In line with the Society of Construction Law Delay *and Disruption Protocol* launched in October 2002, this book also puts forward a rational and sufficiently accurate method of quantifying the effects of disruption in terms of both cost and time.

Disruption claims impact on the whole of the construction industry, so this book is written for all those members of the construction industry who are involved in submitting, evaluating, awarding, managing and resolving disruption claims.

It is my view that the methods used to quantify disruption must be readily usable by site management. Agreement at this level is the target of the solutions proposed, as it is hoped that this prevents the claim escalating to the formal dispute resolution procedures. It has been my experience that resolving claims for delay and disruption at site level reduces the souring of site relationships and prevents loss of senior management/head office time, which in turn prevents the cost of formal dispute resolution (adjudication, arbitration and litigation).

The solutions proposed in this book also seek to be realistic and to recognise that, in practice, any method of quantifying the cost and time effects of delay and disruption must be sufficiently accurate, robust and useful that the method employed at site level can also be used (if needed) by adjudicators, arbitrators and judges.

Construction disputes, albeit nominally about money, invariably involve issues to do with time. Extension of time claims self-evidently involve time, as do claims for Liquidated Damages.

Similarly, claims for prolongation costs, loss and expense or disruption are all fundamentally about time. The effective management of time is therefore a part of everything we do in construction and it is at the heart of all construction contracts.

Practical Guide to Disruption and Productivity Loss on Construction and Engineering Projects, First Edition. Roger Gibson.
© 2015 John Wiley & Sons, Ltd. Published 2015 by John Wiley & Sons, Ltd.

Cost and time are interdependent. From a project management perspective, the treatment of cost (most commonly in the BoQ) and time (in the programme) as independent models fails to provide a mechanism of direct performance/efficiency comparison. It also prevents the systematic evaluation of the effects of variations and delay. Delay and disruption are associated with time and will often have a related impact on cost.

Whilst it may be tempting to require the development of a system that can quantify the costs associated with disruption to almost laboratory standards, it must be remembered that the construction site is not a laboratory and it is simply uneconomical, impractical, unnecessary and unrealistic to expect to develop such a complex system. In practice, there is a need to balance the desire for extreme accuracy with practical reality; – this book recognises this practical hindrance and therefore proposes a solution that is sufficiently accurate for the quantification of disruption claims.

This book aims to demonstrate how the actual site labour productivity measurements can be used to provide an objective and automatic basis for quantifying the effects of disruption in terms of cost and time to arrive at a figure for the loss/expense payable to the contractor. The present position in construction disruption-based disputes is that settlement is often reached after extensive, and sometimes highly subjective, negotiations. The parties' positions are usually severely weakened by a lack of records that may actually demonstrate the effect of a "disruptive" event on the contractor's work operations. If the contractor's productivity could be recorded sufficiently faithfully and simply, it could be used as objective evidence to accurately demonstrate the effect the disruption has actually had on the site productivity. The equating of labour productivity loss to disruption is therefore a realistic and objective measure of the effect of disruption on the contractor's work operation.

Delays are an endemic feature of the construction and engineering industries. In the construction industry, the aim of project control is to ensure the projects finish on time, within budget and achieve other project objectives. It is a complex task undertaken by project managers in practice, which involves constantly measuring progress, evaluating plans, and taking corrective actions when required. During the last few decades, numerous project control methods, such as Gantt Bar Chart, Program Evaluation and Review Technique (PERT) and Critical Path Method (CPM), have been developed. A variety of software packages have become available to support the application of these project control methods, for example Microsoft Project, Asta Power Project, Primavera, etc.

Despite the wide use of these methods and software packages in practice, many construction and engineering projects still suffer time and cost overruns.

There have been numerous studies on the identification of influencing factors of project time and cost overruns worldwide. These studies have found that the most important variables causing construction delays and disruption are: poor contract management; financing and payment of completed works; changes in site conditions; shortage of materials; imported materials and plant items; design changes; and subcontractors.

1.2 The aims of this book

The aim of this book is to provide guidance in relation to disruption and loss of productivity claims. The contents of this book are intended to give its readers the information and practical details to be considered in formulating disruption and loss of productivity claims.

One of the recurring themes in this book is good record keeping on projects. Whilst a lack of progress related records may not be fatal to a claim, it does make a reasonable settlement into an uphill battle. Readers will observe my continuing advice on good record keeping.

The book has been arranged in six chapters:

Chapter 1 Introduction: details general principles relating to extensions of time, delay claims and the SCL Protocol.

Chapter 2 Contracts and Case Law: looks at the relevant loss and expense clauses in the JCT and NEC contracts, plus case law concerning disruption, loss of productivity, mitigation and acceleration.

Chapter 3 Programmes and Record Keeping': deals with the fundamental matter of the project programme, together with the important matter of record keeping during the project.

Chapter 4 Delay and Disruption: looks at the fundamentals of these two issues.

Chapter 5 Loss of Productivity: included in this chapter are examples of two techniques to demonstrate disruption and productivity loss.

Chapter 6 Acceleration and Mitigation: this final chapter looks at the fundamentals of these two issues.

At the end of the book are three appendices which I consider the reader will find helpful. These are:

Appendix 1: Definitions and Glossary.
Appendix 2: Standards for the Levels of a Programme or Schedule.

1.3 Appendix 3: Society of Construction Law: Delay & Disruption Protocol (October 2002) The SCL Protocol

In October 2002, the Society of Construction Law (SCL) published its 'Delay & Disruption Protocol'. This Protocol provides guidance to people dealing with submissions for extension of time and delay claims, both during a contract and after completion of the works. The Protocol runs to some 82 pages and was drafted by a group of experts from all sections of the construction industry.

The Protocol envisages that decision-takers, e.g. contract administrators, adjudicators, dispute review boards, arbitrators, judges, may find it helpful in dealing with time-related issues.

There are 21 Core Statements of Principle in the Protocol. Of these, about 11 relate to 'disruption', 'loss of productivity' and/or 'acceleration'. These are:

1. **Programme and records**: To reduce the number of disputes relating to delay, the Contractor should prepare and the Contract Administrator (CA) should accept a properly prepared programme showing the manner and sequence in which the Contractor plans to carry out the works. The programme should be updated to record actual progress and any extensions of time (EOTs) granted. If this is done, then the programme can be used as a tool for managing change, determining EOTs and periods of time for which compensation may be due. Contracting parties should also reach a clear agreement on the type of records that should be kept.
2. **Concurrent delay – its effect on entitlement to compensation for prolongation**: If the Contractor incurs additional costs that are caused both by Employer Delay and concurrent Contractor Delay, then the Contractor should only recover compensation to the extent it is able to separately identify the additional costs caused by the Employer Delay from those caused by the Contractor Delay. If it would have incurred the additional costs in any event as a result of Contractor Delays, the Contractor will not be entitled to recover those additional costs.
3. **Identification of float and concurrency**: Accurate identification of float and concurrency is only possible with the benefit of a proper programme, properly updated.
4. **Mitigation of delay and mitigation of loss**: The Contractor has a general duty to mitigate the effect on its works of Employer Risk Events. Subject to express contract wording or agreement to the contrary, the duty to mitigate does not extend to requiring the Contractor to add extra resources or to work outside its planned working hours. The Contractor's duty to mitigate its loss has two aspects – first, the Contractor must take reasonable steps to minimise its loss; and secondly, the Contractor must not take unreasonable steps that increase its loss.
5. **Valuation of variations**: Where practicable, the total likely effect of variations should be pre-agreed between the Employer/CA and the Contractor, to arrive if possible at a fixed price of a variation, to include not only the direct costs (labour, plant and materials) but also the time-related costs, an agreed EOT and the necessary revisions to the programme.
6. **Basis of calculation of compensation for prolongation**: Unless expressly provided for otherwise (e.g. by evaluation based on contract rates), compensation for prolongation should not be paid for anything other than

work actually done, time actually taken up or loss and/or expense actually suffered. In other words, the compensation for prolongation caused other than by variations is based on the actual additional cost incurred by the Contractor. The objective is to put the Contractor in the same financial position it would have been if the Employer Risk Event had not occurred.

7. **Relevance of tender allowances**: The tender allowances have limited relevance for the evaluation of the costs of prolongation and disruption caused by breach of contract or any other cause that requires the evaluation of additional costs.
8. **Period of evaluation of compensation**: Once it is established that compensation for prolongation is due, the evaluation of the sum due is made by reference to the period when the effect of the Employer Risk Event was felt, not by reference to the extended period at the end of the contract.
9. **Global claims**: The not uncommon practice of contractors making composite or global claims without substantiating cause and effect is discouraged by the Protocol and rarely accepted by the courts.
10. **Acceleration**: Where the contract provides for acceleration, payment for the acceleration should be based on the terms of the contract. Where the contract does not provide for acceleration but the Contractor and the Employer agree that accelerative measures should be undertaken, the basis of payment should be agreed before the acceleration is commenced. It is not recommended that a claim for so-called constructive acceleration be made. Instead, prior to any acceleration measures, steps should be taken by either party to have the dispute or difference about entitlement to EOT resolved in accordance with the dispute resolution procedures applicable to the contract.
11. **Disruption**: Disruption (as distinct from delay) is disturbance, hindrance or interruption to a Contractor's normal working methods, resulting in lower efficiency. If caused by the Employer, it may give rise to a right to compensation either under the contract or as a breach of contract.

Further background and guidance on each of the 21 Core Principles is contained in the four 'Guidance Sections', which are:

Section 1: Guidelines on the Protocol's position on Core Principles and on other matters relating to delay and compensation.
Section 2: Guidelines on preparing and maintaining programmes and records.
Section 3: Guidelines on dealing with extensions of time during the course of the project.
Section 4: Guidelines on dealing with disputed extension of time issues after completion of the project – retrospective delay analysis.

1.3.1 Observations

Firstly, observations on the Core Principles.

(i) Core Principles 2 to 6: Extensions of time
The position on extensions of time is generally good and the advice is sound, although fairly general in nature.

(ii) Core Principle 7: Float, as it relates to time
This is one of the more controversial Principles in the Protocol. The nub of this principle is
   (a) Should the contractor be awarded an extension of time and so preserve the float period for its own use,
   or
   (b) Should no extension of time be awarded on the basis that the employer's delay is simply absorbing float and not impacting the contractual completion date.
The Protocol's recommendation is that float is available to the project. In other words, it is available to whichever party uses it first: contractor or employer.

(iii) Core Principle 8: Float, as it relates to compensation
Where a contractor plans to complete before the contract date for completion, the Protocol recommends that he is entitled to compensation, but not an extension of time, if he is prevented from completing to his own planned date, but finishes before the contract date for completion. This is a complicated topic; however, the basic recommendation must be rejected. The position is that, in deciding this question, all the circumstances must be taken into account.

(iv) Core Principle 9: Concurrent delay – its effect on entitlement to extension of time
The Protocol's approach seems to be to take a particular position on the subject of concurrency on the basis that it is a complex topic and a compromise situation is necessary. A basic principle is that no concurrent cause of delay which is the result of any fault of the contractor should reduce the extension of time to which he would otherwise be entitled. This approach basically follows the 'prevention principle' of English law where an employer cannot take advantage of its own breach of contract by imposing liquidated damages on the contractor.

(v) Core Principle 13: Mitigation of delay and mitigation of loss
This is a clear exposition of the situation. More could have been said in the Protocol about the contractor's rights, or otherwise, to claim reasonable costs of mitigation.

(vi) **Core Principle 15: Valuation of variations**
The Protocol recommends a mechanism similar to the current JCT price statement for dealing with the valuation of variations and associated extension of time and loss and expense.

(vii) **Core Principle 16: Basis of calculation of compensation for prolongation**
It is rightly stressed that ascertainment must be based on actual additional costs incurred by the contractor. However, there appears to be some confusion between a contractor's claims for loss and expense under the contract machinery and claims for damages for breaches of contract. The former are reimbursable under most standard forms of contracts while the latter, being a claim outside the contract, are not so reimbursable.

(viii) **Core Principle 17: Relevance of tender allowances**
It is refreshing to see that the Protocol considers that tender allowances have little or no reliance to the evaluation of the costs of prolongation or disruption.

(ix) **Core Principle 19: Global claims**
It is good to see that global claims are discouraged.

(x) **Core Principle 20: Acceleration**
This is a broadly correct interpretation of the position, but the reference to the possibility of accelerating by instructions about hours of working and sequence of working is to be doubted.

(xi) **Core Principle 21: Disruption**
The definition of disruption does not adequately explain that disruption can also refer to a delay to an individual activity not on the critical path where there is no resultant delay to the date for completion. The Protocol also states that most standard forms do not expressly deal with disruption; that, of course, is true. However, the JCT forms refer to regular progress being materially affected. That appears to be broad enough to encompass both disruption and prolongation.

The Protocol's Guidance Section 2 deals with guidelines on preparing and maintaining programmes and records. However, there is not a great deal of guidance on maintaining records generally.

Stress is placed on obtaining an 'Accepted Programme'; that is, a programme agreed by all parties. There are several problems with this. Perhaps the foremost is that the architect will be unlikely to have the requisite skills and/or experience, or indeed the information required, to accept the contractor's programme. He is probably capable of questioning parts of it, but highly unlikely to be possessed of sufficient information to be able to satisfy himself that the programme is workable. The Protocol, rightly, accepts that the contractor is entitled to construct the building in whatever manner and sequence he pleases, subject to any sectional completion or other constraints. The Protocol states,

'Acceptance by the CA (contract administrator) merely constitutes an acknowledgement by the CA that the Accepted Programme represents a contractually compliant, realistic and achievable depiction of the Contractor's intended sequence and timing of construction of the works'.

This is placing a responsibility on the architect (or CA as the Protocol prefers) which he is not required to carry. There appears to be no need for a programme to be accepted. It is sufficient if the contractor puts it forward as the programme to which he intends to work. The architect is entitled to question any part which appears to be clearly wrong or unworkable. But, in the light of the contractor's insistence that he can and will carry out the works in accordance with the submitted programme, it is difficult to refuse a programme unless firm objections can be raised.

The Protocol also recommends that the 'accepted programme' be updated with progress at intervals of one month, and more frequently on complex projects.

The Protocol describes the updating process as follows:

'Using the agreed project planning software, the Contractor should enter the actual progress on the Accepted Programme as it proceeds with the works, to create the Updated Programme. Actual progress should be recorded by means of actual start and actual finish dates for activities, together with percentage completion of currently incomplete activities and/or the extent of remaining activity durations. Any periods of suspension of an activity should be noted in the Updated Programme. The monthly updates should be archived as separate electronic files and the saved monthly versions of the Updated Programme should be copied electronically to the CA, along with a report describing all modifications made to activity durations or logic of the programme. The purpose of saving monthly versions of the programme is to provide good contemporaneous evidence of what happened on the project, in case of dispute.'

All of this is good and sensible advice.

Guidance Section 3 gives guidelines for dealing with extensions of time during the course of the project. It provides much good practical advice, including the importance of calculating extensions of time by means of various programming techniques. Although every architect should be familiar with such techniques, careful consideration should be given to the aptness of any particular technique in a given situation.

The Protocol suggests that extensions of time should be made as close in time to the delaying event, and that these are dealt with promptly by the CA. The Protocol recommends that:

'...the "Updated Programme" should be the primary tool used to guide the CA in determining the amount of the EOT.'

Again sound advice, with one proviso: the facts surrounding the alleged delay event(s). As Mr Justice Dyson noted in his judgment on the Henry Boot Construction –v– Malmaison Hotel case: "*It seems to me that it is a question of fact in any case, as to whether a relevant event has caused, or is likely to cause, delay to the works beyond the completion date.*"

Guidance Section 4 deals with disputed extensions of time after completion of the project, and spends some time examining the different types of analysis that can be employed.

1.4 Conclusion

The Protocol sets out ways of dealing with delays and disruption. Most of it is in line with what is generally understood to be the law on these matters. However, in some instances, the Protocol steps outside this boundary in order to suggest what it clearly considers to be a simpler or fairer way of dealing with the practicalities. All parties involved in construction contracts must be aware that the Protocol does not take precedence over the particular contract in use unless it is expressly so stated in the contract itself. Therefore, the Protocol's recommendations should be viewed with caution. It will be of no avail for the architect, contract administrator or employer to argue that he has acted strictly in accordance with the Protocol if the contract prescribes action of a different sort.

And finally...

The author hopes that this book will provide useful guidance for those responsible for preparing extension of time submissions and time-related delay claims; and equally for those people dealing with them. The aim is that these submissions can be resolved amicably, professionally, and with neither party being seriously disadvantaged.

Chapter 2
Contracts and Case Law

2.1 Introduction

This chapter covers two subjects: 'Contracts' and 'Case Law'. The first section concerns the two most popular forms of contract used in the UK: the JCT and the NEC. The chapter reviews the clauses relating to disruption, loss of productivity, acceleration and mitigation.

In the second section, I review case law regarding disputes and judgments concerning disruption, loss of productivity, acceleration and mitigation.

2.1.1 Contracts

All of the standard forms of contract and subcontract include clauses for dealing with delays to the project. Most of these standard forms require the contract administrator (architect or engineer) to deal with a contractor's claim for loss and expense.

For example, JCT2011 contains the following in clause 2.19:

"If and whenever it becomes reasonably apparent that the progress of the Works or any Section is being or is likely to be delayed the Contractor shall forthwith give the Architect/Contract Administrator notice of the cause of the delay. If in the Architect/Contract Administrator's opinion completion of the Works or Section has been, is being or is likely to be delayed beyond the relevant Completion Date by any of the Relevant Events, then, save where these Conditions expressly provide otherwise, the Architect/Contract Administrator, as soon as he is able to estimate the length of the delay beyond that date, shall by notice to the Contractor give a fair and reasonable extension of time for completion of the Works or Section."

2.1.2 Case law

My review of 'case law' in respect of disruption, productivity, acceleration and mitigation covered the cases in the Technology and Construction Court in the period up to the end of 2013.

From this initial review, I examined in detail some 25 disputes which I considered to be the most important concerning disruption and/or loss of productivity, primarily due to the Court's comments and advice on the approach and methodology used. The cases were:

1. Norwest Holst Construction Ltd v. Co-Operative Wholesale Society Ltd [1997] EWHC Technology 356 (2 December 1997)
2. P&O Developments Ltd v. The Guy's and St Thomas' National Health Service Trust, Austen Associates (a firm), Austen Associates Ltd [1998] EWHC Technology 295 (15 October 1998)
3. How Engineering Services Ltd v. Lindner Ceilings Floors [1999] EWHC B7 (TCC) (24 June 1999)
4. Miller Construction Ltd v. James Moore Earthmoving [2000] EWHC Technology 52 (1 November 2000)
5. Royal Brompton Hospital NHS Trust v. Frederick A Hammond & Ors [2000] EWHC Technology 39 (18 December 2000)
6. AMEC Process and Energy Ltd v. Stork Engineers & Contractors BV (No. 3) [2002] EWHC B1 (TCC) (15 March 2002)
7. Johnson Control Systems Ltd. v. Techni-Track Europa Ltd [2002] EWHC 1613 (TCC) (2 August 2002)
8. Royal Brompton Hospital National Health Service Trust v. Hammond & Ors [2002] EWHC 2037 (TCC) (11 October 2002)
9. Joinery Plus Ltd (In Administration) v. Laing Ltd [2003] EWHC 3513 (TCC) (16 January 2003)
10. Home of Homes Ltd v. London Borough of Hammersmith & Fulham & Anor [2003] EWHC 807 (TCC) (10 April 2003)
11. Hurst Stores and Interiors Ltd v. M.L. Europe Property Ltd [2003] EWHC 1650 (TCC) (25 June 2003)
12. Skanska Construction UK Ltd v. Egger (Barony) Ltd [2004] EWHC 1748 (TCC) (30 July 2004)
13. Great Eastern Hotel Company Ltd v. John Laing Construction Ltd & Anor [2005] EWHC 181 (TCC) (24 February 2005)
14. Plymouth & South West Co-Operative Society Ltd v. Architecture, Structure & Management Ltd [2006] EWHC 5 (TCC) (10 January 2006)
15. Multiplex Constructions (UK) Ltd v. Cleveland Bridge UK Ltd [2006] EWHC 1341 (TCC) (5 June 2006)
16. Mirant Asia-Pacific Construction (Hong Kong) Ltd v. Ove Arup and Partners International Ltd & Anor [2007] EWHC 918 (TCC) (20 April 2007)

17. VGC Construction Ltd v. Jackson Civil Engineering Ltd [2008] EWHC 2082 (TCC) (15 August 2008)
18. Fitzroy Robinson Ltd v. Mentmore Towers Ltd [2009] EWHC 1552 (TCC) (7 July 2009)
19. BSkyb Ltd & Anor v. HP Enterprise Services UK Ltd & Anor (Rev 1) [2010] EWHC 86 (TCC) (26 January 2010)
20. Tinseltime Ltd. V. Eryl Roberts [2011] EWHC 1199 (TCC) (13 May 2011)
21. Carillion JM Ltd v. Phi Group Ltd [2011] EWHC 1379 (TCC) (15 June 2011)
22. Carillion Construction Ltd V. Stephen Andrew Smith [2011] EWHC 2910 (TCC) (10 November 2011)
23. Walter Lilly & Company Ltd v. Mackay & Anor [2012] EWHC 1773 (TCC) (11 July 2012)
24. Constance Long Term Holdings Ltd v. Cavendish (Duke of Westminster) [2012] EWHC 3434 (TCC) (29 November 2012)
25. Cleveland Bridge UK Ltd v. Severfield - Rowen Structures Ltd [2012] EWHC 3652 (TCC) (21 December 2012)

From my detailed review of these 25 judgments, I selected three in which I consider the Court made significant observations and gave meaningful advice in respect of *'disruption', 'productivity', 'acceleration'* and *'mitigation'*. The three cases are:

1. AMEC Process and Energy Ltd v. Stork Engineers & Contractors BV (No. 3) [2002] EWHC B1 (TCC) (15 March 2002)
2. Johnson Control Systems Ltd v. Techni-Track Europa Ltd [2002] EWHC 1613 (TCC) (2 August 2002)
3. Cleveland Bridge UK Ltd v. Severfield - Rowen Structures Ltd [2012] EWHC 3652 (TCC) (21 December 2012)

For each of these disputes I give details of (i) the project concerned, (ii) the dispute itself, (iii) the judgment and, finally (iv) a commentary on the important comments and advice given by the Court.

2.2 Contracts

All of the standard forms of contract and subcontract include clauses for dealing with delays to the project, and require the contract administrator (architect or engineer) to deal with a contractor's claim for loss and expense.

The JCT205, Intermediate Building Contract, contains the following two clauses under *'Adjustment of Completion Date'*:

Notice of delay – extensions
Clause 2·19

1. If and whenever it becomes reasonably apparent that the progress of the Works or any Section is being or is likely to be delayed the Contractor shall forthwith give the Architect/Contract Administrator notice of the cause of the delay. If in the Architect/Contract Administrator's opinion completion of the Works or Section has been, is being or is likely to be delayed beyond the relevant Completion Date by any of the Relevant Events, then, save where these Conditions expressly provide otherwise, the Architect/Contract Administrator, as soon as he is able to estimate the length of the delay beyond that date, shall by notice to the Contractor give a fair and reasonable extension of time for completion of the Works or Section.
2. If any Relevant Event referred to in clauses 2·20·1 to 2·20·6 occurs after the relevant Completion Date but before practical completion is achieved, the Architect/ ContractAdministrator, as soon as he is able to estimate the length of the delay, if any, to the Works or any Section resulting from that event, shall by notice give a fair and reasonable extension of time for completion of the Works or Section.
3. At any time up to 12 weeks after the date of practical completion of the Works or Section, the Architect/Contract Administrator may give an extension of time in accordance with the provisions of this clause 2·19, whether on reviewing a previous decision or otherwise and whether or not the Contractor has given notice as referred to in clause 2·19·1. Such an extension of time shall not reduce any extension previously given.
4. Provided always that the Contractor shall:
    1. constantly use his best endeavours to prevent delay and do all that may reasonably be required to the satisfaction of the Architect/Contract Administrator to proceed with the Works or Section; and
    2. provide such information required by the Architect/Contract Administrator as is reasonably necessary for the purposes of this clause 2·19.
5. In this clause 2·19 and, so far as relevant, in the other clauses of these Conditions, any reference to delay or extension of time includes any further delay or further extension of time.

Relevant Events
Clause 2·20
The following are the Relevant Events referred to in clause 2·19:

1. Variations and any other matters or instructions which under these Conditions are to be treated as, or as requiring, a Variation;
2. Architect/Contract Administrator's instructions:
    1. under any of clauses 2·13, 3·12 or 3·13 (excluding, where there are Contract Bills, an instruction for expenditure of a Provisional Sum for defined work);
    2. (to the extent provided therein) under clause 3·7 and Schedule 2; or

3. for the opening up for inspection or testing of any work, materials or goods under clause 3·14 or 3·15·1 (including making good), unless the inspection or test shows that the work, materials or goods are not in accordance with this Contract;
3. deferment of the giving of possession of the site or any Section under clause 2·5;
4. the execution of work for which an Approximate Quantity is not a reasonably accurate forecast of the quantity of work required;
5. suspension by the Contractor under clause 4·11 of the performance of his obligations under this Contract;
6. any impediment, prevention or default, whether by act or omission, by the Employer, the Architect/Contract Administrator, the Quantity Surveyor or any of the Employer's Persons, except to the extent caused or contributed to by any default, whether by act or omission, of the Contractor or of any of the Contractor's Persons;
7. the carrying out by a Statutory Undertaker of work in pursuance of its statutory obligations in relation to the Works, or the failure to carry out such work;
8. exceptionally adverse weather conditions;
9. loss or damage occasioned by any of the Specified Perils
10. civil commotion or the use or threat of terrorism andlor the activities of the relevant authorities in dealing with such event or threat;
11. strike, lock-out or local combination of workmen affecting any of the trades employed upon the Works or any of the trades engaged in the preparation, manufacture or transportation of any of the goods or materials required for the Works;
12. the exercise after the Base Date by the United Kingdom Government of any statutory power which directly affects the execution of the Works;
13. *force majeure.*

Also included in the JCT205, Intermediate Building Contract, are the following two clauses under '*Loss and Expense*':

Disturbance of Regular Progress
Clause 4·17
If in the execution of this Contract the Contractor incurs or is likely to incur direct loss and/or expense for which he would not be reimbursed by a payment under any other provision in these Conditions due to:

1. a deferment of giving possession of the site or relevant part of it under clause 2·5; or
2. the regular progress of the Works or any part of them being materially affected by any of the Relevant Matters,

and the Contractor makes an application to the Architect/Contract Administrator within a reasonable time of that becoming apparent, then, save where these Conditions provide that there shall be no addition to the Contract Sum or otherwise exclude the operation of this clause, the Architect/Contract Administrator, if and as soon as he is of that opinion, shall ascertain, or instruct the Quantity Surveyor to ascertain, the amount of the loss and/or expense incurred and that amount shall be added to the Contract Sum, provided that the Contractor shall in support of his application submit such information required by the Architect/Contract Administrator or the Quantity Surveyor as is reasonably necessary for the purposes of this clause 4·17.

Relevant Matters
Clause 4·18
The following are the Relevant Matters:

1. Variations (including any other matters or instructions which under these Conditions are to be treated as, or as requiring, a Variation);
2. Architect/Contract Administrator's instructions:
   1. under clause 3·12 or 3·13 (excluding, where there are Contract Bills, an instruction for expenditure of a Provisional Sum for defined work);
   2. (to the extent provided therein) under clause 3·7 and Schedule 2;
   3. for the opening up for inspection or testing of any work, materials or goods under clause 3·14 (including making good), unless the inspection or test shows that the work, materials or goods are not in accordance with this Contract; or in relation to errors, omissions and inconsistencies in or between the Contract Documents and/or other documents referred to in clause 2·13;
3. suspension by the Contractor under clause 4·11 of the performance of his obligations under this
4. the execution of work for which an Approximate Quantity is not a reasonably accurate forecast of the quantity of work required;
5. any impediment, prevention or default, whether by act or omission, by the Employer, the Architect/Contract Administrator, the Quantity Surveyor or any of the Employer's Persons, except to the extent caused or contributed to by any default, whether by act or omission, of the Contractor or of any of the Contractor's Persons.

2.2.1 The main changes in JCT2005 affecting EOT submissions

The main difference in the wording between the earlier forms is that JCT2005 does not include any reference to 'nominated subcontractors'; and introduces two new terms: 'Pre-agreed Adjustment' and 'Relevant Omission'.

The approach taken by the 1998 and 2005 contracts is generally the same; the contractor has to notify the contract administrator of the cause or causes of delay,

"...whenever it becomes reasonably apparent that the progress of the Works or any Section is being or is likely to be delayed."

The contract administrator then has 12 weeks to give an extension of time and fix a new completion date with adjustments for relevant events, which are listed in clause 2·20.

The 12-week period was introduced in the 1980 version of the JCT contracts in order to introduce a limit to the amount of time the contract administrator or architect could take for an extension of time decision. Many observers were of the opinion that one of the principle reasons there was a slow take-up of the 1980 contract was that it was considered unreasonable that there should be a time limit on the architect's decision-making process. However, there has always existed a 'get-out' condition in that the contract administrator had only to deal with EOT submissions when it had received 'reasonably sufficient particulars'.

JCT2005 has tightened the rules on how contractors and contract administrators must approach extensions of time. For example, the qualification that the contract administrator has to have 'reasonably sufficient information' in order to be able to make a decision has been deleted. This appears to mean that the contract administrator must make a decision even on insufficient information. However, the contract administrator needs to make his decision, and fix a new date which it considers fair and reasonable, and this will be based on the information received.

Another example is that the contractor has to give a wider picture of all the delays and the reasons for them, and the contract administrator has to respond to all of the events and say in detail what it is giving extra time for.

Other changes affecting extensions of time are:

1. The contractor has to give notice of a relevant event and estimate 'the expected delay in the completion of the works'. JCT1998 went on to say 'whether or not concurrently with delay resulting from any other relevant event'. This has been omitted from JCT2005. However, although that wording has been left out, the contractor's obligation is still to provide an estimate of the delay in relation to any relevant event, regardless of whether they are concurrent with other events.
2. The contract administrator has to 'decide' on extensions of time, whereas under JCT1998 he had 'in writing to the contractor give an extension of time'. This change in words indicates a more active approach.
3. The contract administrator's decision has to state the extension that has been attributed to each relevant event. Previously, in JCT1998, the obligation was that the decision only had to state which of the relevant events had been taken into account.
4. If the contract administrator's decision is that no extension of time is due then there is an obligation to notify the contractor so, in writing. In JCT1998 there appeared to be no obligation for the contract administrator to respond to the contractor if an extension of time had not been granted.

5. Clause 25·4·10 in JCT1988: the unavailability of labour and materials. This sub-clause has been deleted in JCT2005. As this was invariably deleted in the previous version, the omission may not make much difference.
6. Where the period of time between the particulars being received and the completion date is less than 12 weeks, then the contract administrator has only to 'endeavour' to provide its decision prior to the completion date. Previously in JCT1998, the obligation was to make a decision 'if reasonably practicable'.

2.3 Case law

My review of 'case law' is in respect of disruption, productivity, acceleration and mitigation covered the cases in the Technology and Construction Court in the period up to the end of 2013.

From this initial review, I examined in detail some 25 disputes which I considered to be the most important in concerning disruption and/or loss of productivity, primarily due to the Court's comments and advice on the approach and methodology used.

From my detailed review of these 25- judgments, I selected three in which I consider the Court made significant observations and gave meaningful advice in respect of *'disruption', 'productivity', 'acceleration'* and *'mitigation'*. The three 'cases' are:

1. AMEC Process and Energy Ltd v. Stork Engineers & Contractors BV (No. 3) [2002] EWHC B1 (TCC) (15 March 2002)
2. Johnson Control Systems Ltd v. Techni-Track Europa Ltd [2002] EWHC 1613 (TCC) (02 August 2002)
3. Cleveland Bridge UK Ltd v. Severfield - Rowen Structures Ltd [2012] EWHC 3652 (TCC) (21 December 2012)

For each of these disputes I give details of (i) the project concerned, (ii) the dispute itself, (iii) the judgment and, finally (iv) a commentary on the important comments and advice given by the Court.

The quantification of reasonable costs in respect of particular subjects that are covered by the judgments included in this chapter are: (i) 'Disruption'; (ii) 'Productivity'; (iii) 'Acceleration' and (iv) 'Mitigation'.

2.3.1 AMEC Process and Energy Ltd v. Stork Engineers & Contractors BV (2002)

2.3.1.1 The Facts

In 1995 Shell UK Ltd engaged a Swiss company called Single Buoy Moorings Inc ('SBM') to provide a floating production unit for the extraction and processing of crude oil from the North Sea. Although it was to work at a fixed location it was based on a ship's hull and had a ship's name, 'The Anasuria'. SBM procured the hull from a Japanese shipyard (Mitsubishi) and engaged Stork to design, fabricate, construct, install and precommission the topside facilities. Stork carried out the design itself and supplied certain materials but subcontracted the remainder of its contract works to 'Amec'.

2.3.1.2 The Dispute

The early history of the claims may be briefly summarised. The topsides were being designed, fabricated, constructed and installed on the barge between 15 February 1995 when the contract took effect and 14 June 1996. Additional work was undertaken on the barge until sailaway on 24 August 1996. From November 1995, Amec submitted claims for the disruption and acceleration measures that were occurring in the form of both individual claims and monthly valuation applications. With few and limited exceptions, Stork rejected these various applications and claims and little was certified or paid towards them. During the later months of 1996 and up to the last commercial meeting held on 20 February 1997, limited and spasmodic commercial meetings and negotiations took place but little was agreed or paid.

In the next six months, Amec's claims were then finalised and submitted to Stork. These were submitted initially in the form of a final interim application for payment dated 13 March 1997 and then in the form of claims submissions based on taking three sample modules as examples of the overall loss that had occurred. These claims were the subject of a high-level claims meeting and claims negotiations held in Schiedam on 5 July 1997. This meeting was agreed to be subject to without-prejudice privilege. The breakdown of these negotiations at a further without-prejudice meeting held in Schiedam on 24 September 1997 led to Amec working on the details of its pleaded claim from October 1997 until the statement of claim was served in June 1998.

The striking feature of all of Amec's applications for payment, valuations submissions, claims submissions and supporting details from the first moment when they were prepared by Amec in about November 1995, midway through the contract, until the lead-up to the two Schiedam meetings, is that all of them contained details of the claims which were ultimately tried. The claims evidenced by all these documents were based on the lost hours for which Amec contended there had been no payment. Moreover, all these claims were based on the same variations and late submissions of non-conforming design drawings that Amec relied on at the trial. Finally, all these claims were seeking to recover Amec's costs and losses arising from the delay, disruption and acceleration that these variations and drawing submissions had caused.

By the conclusion of the second Schiedam meeting, Amec was in no doubt of Stork's overall approach to its claims. Stork's approach was, in summary, that:

- no disruption had occurred;
- no link between any variation or late issued drawing and any disruption had been established;
- the claims failed for being presented as global claims and that the only approach to the verification of loss would be one which linked lost hours associated with individual drawings and variations to specific hours and employee's time;
- disruption could only be seen to have occurred if employees were unable to work at all rather than being able to work unproductively;
- the sampling method of seeking to establish claims was inherently flawed;
- any loss that occurred was occasioned by Amec by the inadequate rates contained in its tender and by its working methods; and
- the contract conditions precluded the claims both because the suggested loss was defined as being included within the contract rates and because of repeated failures to follow the procedural conditions precedent to payment.

In 1998, Amec served a statement of claim with adjustments of price or damages to a total of some £13.75 M, and Stork counterclaims liquidated damages, price reductions and repayment of overpayments to a total of some £2.85 M.

The action resulted in a series of trials taking place in the Technology and Construction Court between 1999 and 2002.

Trial Number 1

Before His Honour Judge Hicks QC: Judgment issued on 6 May 1999.

The Judgment

By his Order for Directions dated 16 October 1998 Judge Humphrey Lloyd QC directed the trial of certain issues, expressed in general terms which I need not recite, since they were to be, and were, more precisely defined by agreement between counsel and since then have, again by agreement, been revised. Finally, the parties have come to agreed terms which dispose of some of those revised issues, leaving only those numbered 1, 3, 6 and 8 to be tried. There is also an agreed "position paper" which helpfully explains the background to those remaining issues. Since each largely turns on a discrete set of contractual provisions and is expressed in terms which presuppose some acquaintance with its contractual and factual setting it is best to set each out separately, after describing that setting and immediately before dealing with any disputed evidence and the submissions on that issue.

I now focus on 'Issue 6 – Acceleration'. The Judgment contained:

In clause 13·1 the following passages are particularly material:
(a) [Stork] shall have the right at any time to issue instructions to [Amec] to do and [Amec] shall do any of the following:
   (iv) Accelerate the WORK within limits of practicality in order to recover all or any part of the delay in respect of which [Amec] would otherwise have been entitled to a revision to the PROTECH PLAN in accordance with clause 5 below.
   (v) Reprogramme the WORK and reschedule its resources within the limits of practicality in order to complete the WORK or any part thereof in accordance with any amendment to the PROTECH PLAN which [Stork] may require.

Sub-clause (b) deals with adjustments to the Protech Plan as follows:
(b) .... A fair and reasonable adjustment to the PROTECH PLAN shall be made taking into account all relevant factors including any acceleration instructed under sub-clause 1(a)(iv) above.

Against that background issue 6 asks:
Is Amec entitled to sums claimed to have been incurred by it in respect of accelerative measures:
(a) Pursuant to Article 13; and/or
(b) Otherwise than pursuant to Article 13;

There is a final difficulty, which concerns the meaning of the words "accelerative measures". "Acceleration", as an ordinary English word, means simply "increase in speed". In the context of this contract that would entail finishing, or possibly reaching some intermediate stage, before the contract date. I did not gain the impression that that was the basis of many of Amec's claims, or perhaps of any. There is, however, a clear exposition in clause 13.1(a)(iv) itself of what "accelerate", as used in that sub-clause, means. It involves taking steps "in order to recover .... delay in respect of which [Amec] would otherwise have been entitled to a revision to the PROTECH PLAN". It does not, by inference, include such reprogramming of work and rescheduling of resources as is dealt with separately in the immediately following category (v).

Judging by the examples canvassed in argument, however, the sums which Amec seeks to claim "in respect of accelerative measures" may cover a much wider field. They seem to include measures within category (v) as well as category (iv) of sub-clause (a), the increase of resources generally in order to cope with increases in the quantity of work, measures taken in consequence of breaches of contract by Stork, or in order to mitigate those consequences and, most generally, any use of "additional resources" in consequence of Stork's instructions, acts or omissions. "Additional" implies some basis of comparison - additional to what? - and it was not clear to me whether claims of that kind were advanced on the basis of there being some contractually specified level of resources or whether the comparison was simply with Amec's internal pre-planning.

In the circumstances described in paragraphs 99 to 102 above I do not think it possible to answer this issue in the terms in which it is posed, nor do I find it helpful to try to re-word it. In particular I believe it best to avoid the use of "acceleration" and its derivatives in any undefined sense. I do not for myself, however, find the principles which should be applied to the issues canvassed and examples advanced under this head particularly obscure or much open to debate, and I shall endeavour to state them in the following paragraphs.

(c) In either case was an express instruction (whether oral or in writing) a pre-requisite to recovery of such sums?

The amount of any entitlement under paragraph 104 or 105 above will be that provided for by clause 13·5(a). Unless the variation in question is expressly an "acceleration" within the meaning of clause 13·1(a)(iv) the question whether that entitlement includes the cost of what Amec may classify as "accelerative measures" cannot be answered in any general way; it is one to be decided by the application of the pricing provisions of clause 13·5(a) to the facts of the case.

Where Amec is entitled either to an extension of time or alternatively, in lieu of such an extension, to an instruction to accelerate in accordance with clause 13·1(a) (iv), Stork has the option which to award, without prejudice to any claim for damages which Amec may have if Stork fails to comply with its duties in dealing with the matter, for example under any express or implied time limits within which it must reach or communicate decisions. This is the first of the two instances in which the propositions which I am setting out may not be essentially common ground, but in my view the existence of such an option is properly to be deduced from Article 13 as a whole, and specifically from clauses 13·1(a)(iv), 13·4(a), 13·5(b) and 13·6(i).

Amec is entitled to the benefit of clause 13·6, subject to compliance with the procedural requirements of clause 13·7, and the question whether the cost of what Amec may classify as "accelerative measures" shall be included in any consequential price adjustment is one to be decided by the application of the pricing provisions of clause 13·6 to the facts of the case.

I conclude that the necessary and sufficient criteria for additional payment for allegedly "accelerative measures" were the same after 28 March 1996 as before, namely those explained above in dealing with issue 6. That conclusion entails rejecting Stork's case, if it was still maintained in its full rigour, that clause 2 excludes all liability on its part for the costs of accelerative measures taken after 28 March 1996. So far as such costs are recoverable by those criteria they remain so, whether incurred before or after that date and whether arising from variations authorised before or after that date. It may well be that in practice the criteria produce results somewhat similar to those intended to flow from Amec's contentions, based on distinctions of the kind described in paragraph 117 above, because after 28 March 1996 Stork would clearly not be prepared to grant, and

Amec could hardly request, variations to cover additional costs which Amec was incurring by its own choice with a view to qualifying for the incentive payments. For the reasons which I have given, however, they have two advantages over those distinctions. The first is that they provide straightforward tests instead of difficult, contentious or impossible ones. The second and decisive advantage is that they are what the contract and the incentive agreement, taken together, require instead of being devised and imposed out of thin air.

Trial Number 2

Before His Honour Judge Hicks QC: Judgment issued on 7 December 1999

The Judgment
"By his Order for Directions dated 16 October 1998 Judge Humphrey Lloyd QC directed the trial of a number of preliminary issues in this action. Those issues were tried before me on 16, 17, 18 and 24 March 1999 and I delivered a reserved judgment on 6 May 1999. Although the outcome was not completely one-sided Mr Gray, for the Defendant, concedes that, in his own words, "the substantive victors were the Claimants".

Mr White, for the Claimant, now applies for an order in the Claimant's favour of the costs of and caused by the trial of those issues and for an order for a payment on account of those costs. Mr Gray seeks an order that those costs be reserved until the conclusion of the trial or trials of the remaining issues.

Conclusion
Taking all the relevant circumstances into account, and in particular the considerations canvassed above, I have come to the conclusion that I should exercise my discretion as to costs in this case by reserving the costs of the preliminary issues tried before me to the conclusion of the trial or other disposal of the remaining issues in the action."

Trial Number 3

Before His Honour Judge Thornton QC: Judgment issued on 15 March 2002

The Judgment
Judge Thornton outlined the problems of claim presentation arising from the Wharf Properties case and other cases concerned with disruption claims of a global nature. These being:

"3.4.2. The Problems Confronting the Claims Team
  25. The team were acutely aware of the problems of claim presentation arising from the Wharf Properties case and other cases concerned with disruption claims of a global nature. The team saw the problems as being these:
   1. Disruption, particularly when it had been so extensive and caused by so many different causes, is difficult to establish and a causal link between individual causes and losses almost impossible to show.
   2. Global claims, that is claims which are based on establishing that the totality of disruption losses was caused by the totality of causative events, may in principle be claimed but are likely to fail unless each supporting fact is established in full.
   3. There is no agreed or clearcut method or methods of identifying and quantifying disruption and many of the available methods are widely held to be incapable of satisfying empirical, contractual or evidential standards of proof.
   4. Whatever method of proof is adopted, the pleadings should set out a party's case as to causative events, causation, loss and methods of proof with sufficient clarity so as to enable the opposing party to know what case it is facing, to prepare its own case and to provide a manageable agenda for the trial. In particular, a claiming party should set out what case, if any, it is advancing as to how any global or rolled up claim should be reduced or valued in the event of it achieving only partial success in the proof of the primary facts that it is relying on.
   5. Where the project, as this one, was a large one with many thousands of documents and with many documents in electronic form, the marshalling of the data and its analysis would depend, in part, on the use of appropriate software. Choosing the appropriate programs, adapting the programs and loading the software with the primary data would be difficult, skilled, time consuming and expensive operations and much care and thought would be needed in these exercises.
   6. The preparation of the evidence to support the claims would involve a huge task of assembly and presentation so that it emerged in a reliable and meaningful form. There was, in reality, no obvious dividing line between the pleadings and particulars setting out the claims, the primary evidence needed to establish the claims, the methods of analysis needed to marshal and analyse the evidence and the expert quantity surveying evidence needed to reach conclusions as to the causes and consequences of disruption that had occurred.
   7. At every stage of preparation, Amec would have to tread an almost impossible dividing line between failing to particularise its case sufficiently and providing so much detail that no clearcut agenda for trial had emerged.

> 8. Given Stork's attitude to Amec's claims that had been displayed throughout the lengthy process of applications, discussions, negotiations and settlement talks, it was reasonable to anticipate that Stork would seek to harry Amec procedurally at every step along the road to trial and therefore the procedural steps had to be prepared with unusual care and detail.
> 9. Stork would contest each part of each step along the road to recovery. This contested approach would include Stork disputing the following: the existence of the primary facts relied on; the allegations of causation and loss; the contractual basis of claim; the authenticity of the records relied on; the existence of the suggested procedural preconditions to payment and the suggested methods of establishing the claims by resort to sample modules and lump sum or global methods of proof. Stork would also seek to show that Amec had mismanaged the project in almost all of the steps that it had taken.

The Judgment describes Amec's principal methods of analysing and presenting the supporting data that it regarded as being necessary to establish that a huge number of lost or unremunerated hours had occurred and that virtually all of these lost or unremunerated hours were caused by the factors it relied on.

> (1) Establishing the Contract Ground Rules
> 38. Amec's approach was that it should first establish the contractual ground rules under which the work was carried out. This involved three exercises.
> 39. Firstly, the number of hours that the contract price could be said to have provided for had to be ascertained. Any work carried out within that overall contractual framework was not claimable since it was to be taken to have been included in the contract rates for payment. However, this involved the definition of the contractual scope of work, a difficult task.
> 40. Secondly, the scope of work provided for in the contract price had to be identified. Since the work had only be partly designed at contract stage but Amec had contracted to complete the whole work for a lump sum and within a defined timescale and since the difference between design development and varied work was difficult to identify, this was an important but difficult task.
> 41. Thirdly, Amec had to define the programme or contractual requirement for the delivery of design drawings since no claim could be maintained for any drawings delivered within that contractual timescale. Since the Rev A programme was inevitably rudimentary, this was a difficult task.
>
> (2) Base Programme and As Built Programmes
> 42. Amec's first task was to construct a so-called base programme and as built programmes which identified the notional programme that would have

been issued as Rev A had the full scope of work been identified at the contract stage. This involved a detailed analysis of the tender build up, the scope of work and the variations and the issued drawings. Using this work, Amec constructed both types of programme.

(3) Variations, Drawing Issues, Holds, Materials Issues, Module Work Progress
    43. Amec then had to analyse the work content and work progress of the work. This involved a module by module reconstruction of the drawings issued, the work undertaken and the progress of the work. The issue of materials, the production of spools, the fabrication, installation, shot blasting, painting and module movement through the various stages of production had all to be identified and tabulated.

(4) Hours Worked and Costs
    44. Amec then had to identify the hours actually worked and the costs that had been incurred, largely by reference to manhours, materials and plant usage and overheads.

HHJ Thornton detailed the 'Analytical Methods' used by Amec in their claim presentations, and he noted that Stork never suggested any other or different methods.

"(5) Analytical Methods
    45. There are many different ways of analysing the mass of data assembled by Amec. It chose the following and Stork never suggested any other or different methods:
Comparisons of Programmed and Actual Drawing Delivery.
    As a way of showing, in overall and representational terms, the potential for disruption, Amec compared the pattern of delivery of drawings module by module with the contractual delivery dates provided for by its base programme exercises. This work was much misunderstood by Stork throughout the case. It did not purport to show actual disruption but merely that a project where the actual delivery of necessary design information departed so radically from the intended pattern would inevitably be one which was subject to severe disruption.

Productivity.
    Amec prepared graphs showing a comparison between the output achieved by the labour force, calculated by reference to payments, and the output that would have been achieved had the intended or programmed methods of construction been achieved. These showed a significant reduction in productivity.

Impact Programmes.

Amec prepared impact programmes which attempted to show what progress would have been achieved had the events that occurred not been matched with any increase in resources. This exercise, an adjunct and follow up to the base programme exercise, showed a notional increase in the contract period of at least 10 months would have occurred but for an increase in resources. This was a well-known technique which is often used as a representational means of showing that delay and disruption must have occurred. It was never intended to provide a quantified claim.

Lost Hours.

Amec attempted to determine, using empirical methods of analysis, what number of lost hours had occurred as a result of the separate and discrete heads of claim it had identified. Since these heads of claim were merely artificial parts of the whole, being the total number of lost hours, they were not separate claims but were merely an attempt to make the overall claim both intelligible and one which would not fail if Amec lost on its primary global or Wharf Properties claim.

Breaches and Variations.

Amec also sought to divide the causes of disruption into breaches of contract and variations. This exercise was essentially an insurance policy in case Stork succeeded in showing that contractually based valuation claims were not maintainable or that late delivery of drawings did not amount to a default by Stork. It was also undertaken because there were fears that Amec could not recover loss of profit on its valuation claims. However, since Article 13 succeeded in effectively merging breaches and variations, this exercise was ultimately not needed.

Building Up the Lost Hours.

The empirical approach adopted by Amec involved it in an attempt to use readily useable models or formulae which had been devised empirically by claims consultants over a number of years as a means of showing that disruption must have occurred in defined situations. This approach, albeit a rough and ready one, could provide a formulaic means of determining what degree of disruption had occurred in defined circumstances. This data was ultimately of no assistance to Amec because it was modelled on levels of disruption of a different and lower order of magnitude than that that Amec was subject to."

HHJ Thornton had some significant comments on 'Presentation and Proof', as follows:

"The statement of claim, when served, was the longest and most detailed pleading that I have ever seen. It ran, with the particulars served later, to 12 files and contained a huge number of schedules, appendices and annexes. The document as a whole lacked any supporting commentary which summarised for the uninitiated the structure and contents of the document and a guide to the logic, structure, thinking or method of production. Furthermore, most of the supporting schedules and the like constituted evidence. Although these are cogent criticisms that may be made of the original pleading, Stork never sought to obtain from Amec a supporting commentary at the time the pleading was first served and, had it asked for such a document, Amec would undoubtedly either have been ordered to serve it or would have rapidly produced it without order. Its absence should not now count against Amec when costs are being considered. Indeed, when Stork sought to strike out the statement of claim in its entirety, it did not do so on grounds of prolixity, undue complexity or incoherence but, instead on the substantive ground that the claim as pleaded was global in nature and consequently unsustainable in law. That application was unsuccessful, indeed Judge Lloyd made no order on the application following a two-part contested hearing. Furthermore, I ruled at the trial that the schedules should be decoupled from the pleadings and should stand as evidence, thereby reflecting the reality of these documents. Thus, Amec had served with its pleading a vast quantity of evidence which constituted much of the primary facts needed to support its pleaded case. The criticism that it was served in a somewhat unwieldy form is one of form.

One particular schedule was particularly valuable. This was schedule 24 which contained an indispensable and coherent summary of the history of the contract. It became the working guide at the trial for the factual evidence and enabled all factual to be put into its appropriate context. Indeed, once its content and structure had been mastered, it served as the overall guide to Amec's factual case whose lack I have already drawn attention to. However, much of its content consisted of a summary of the primary evidence which was itself summarised in the plethora of schedules and the like."

The Judgment contained the following, under 'The Claims and Counterclaims':

"Amec's claims in the proceedings that were set out in the statement of claim served on 12 June 1998 were of three separate but related categories which were as follows: (1) a total of £3.3 m was claimed under 24 different heads for variations. (2) A total of £9.46 m was claimed for the combined effects of these heads of claim: (i) prolongation; (ii) disruption; (iii) secondary disruption and (iv) acceleration. Consequent head office overhead expenditure and profit on these heads of claim were also claimed. (3) An extension of time and a consequent release from any liability to pay liquidated damages were claimed.

There were few disputes as to the existence and content of the alleged variations so that these disputes as to variations were largely about their quantification and, in particular, as to the extent that the contract rates for variations should be enhanced so as to give effect to the delayed and disrupted working conditions under which they had been performed. Amec claimed that these conditions had resulted in additional hours being worked which were not provided for in the contract rates. The cause of the second and third categories of claim were pleaded as being the large number of variations, the late supply of drawings and materials and the non-fabrication friendly form in which the design information was provided."

The various headings under which the claim for £9.46 m was claimed were separated out into claims under the contract and claims for damages for breach of contract. These two types of claim were effectively the same since the contract provided that any breach of contract was to be valued as if it had been a variation to the contract. These claims were then further separated out into the 4 headings that I have summarised. The essence of each head of claim was that additional unremunerated hours were worked as a result of the variations and the late issue and unfriendly nature of the many design drawings. These events were also pleaded as having caused the time overrun that had occurred.

Thus, as pleaded, all of Amec's pleaded claims were based on the lost hours that had been worked as a result of the many separate variations and the late issue of drawings that had occurred. These lost hours represented additional hours for which no remuneration had been recovered.

Stork's counterclaims were put forward as both set offs against Amec's claims and as counterclaims standing alone and were of three kinds. Firstly, Stork claimed liquidated damages for delay. Secondly, it claimed a credit or repayment of sums allegedly overpaid because certain items of work had been overvalued. Thirdly, there was an additional category of contra charges. These counterclaims totalled as follows: (1) liquidated damages which totalled approximately £1 m; (2) overpayment and (3) contra charges which together totalled £1.24 m. The liquidated damages claim was intimately interconnected with Amec's claims since the alleged delay in achieving the stages of mechanical completion and being ready for pre commissioning occurred, on Amec's case, entirely as a result of the factors on which it based its claims. The remaining suggested deductions and contra charges were also closely connected with the claims since the facts and events on which the underlying claims were based were the same as those raised by Amec's claims and also by Stork's defences to those claims and, in truth, involved an attack on discrete parts of the overall claim for lost hours and its constituent parts put forward by Amec."

The Judgment states under 'Overall Approach',

"4.1. Claim and Counterclaim

The first question I must consider is whether the action should, for costs purposes, be looked at as a whole or whether, instead, the claim and counterclaim should be considered separately. Stork's overall approach involves a separate consideration of these two parts of the case.

However, I am clear that Stork's counterclaims were so intimately part of the defences it was pursuing and so wrapped up in the claims that Amec was pursuing that it would be nonsensical to treat the counterclaims as separate or discrete from, the claims. The appropriateness of this approach can be seen when it is realised that the counterclaim for liquidated damages was a necessary corollary to Stork's case that the only delay and disruption that occurred was of Amec's sole making and the counterclaim for a revaluation of some of the work items and the counterclaims for defective or inappropriate work were part of Stork's defence and an antidote to Amec's various claims for payment for unremunerated or lost hours.

4.2. Variations and Disruption
A further suggested distinction sought by Stork was between the claims for variations and those for disruption. It is true that the variations claims were separately itemised in discrete schedules and passages in the pleadings, were ordered to be tried separately and were compromised separately from any discussion about other claims and were worked on by different members of the parties' legal and expert witness teams from those working on the balance of the case. However, this separation occurred because of a series of pragmatic decisions taken by the court and the parties as part of their ongoing attempts to make the overall dispute manageable.

The factual reality was that the variations disputes were about disruption and the extent to which Amec could claim an entitlement to the additional costs of executing these variations on top of contract rates because of the additional labour resources needed to execute them compared with the labour resources built into those contract variation rates. The overall settlement of the variations claims just before the first trial, largely to be devoted to these claims, was achieved because the parties, in reality, re-allocated the disruption part of these claims back into the main dispute and then relatively easily, albeit following much expert witness time, settled the remaining parts of the variations disputes.

4.4. The Size of Amec's Recovery
Stork, in its costs submissions, suggested that there had been a substantial lack of recovery by Amec or, in other words, there was a substantial shortfall between sum claimed and recovered. In monetary terms, Amec's claims, in total, were for about £11.5m for which it recovered about £6.4m. By way of comparison, Stork's counterclaims totalled about £2.4m for which it recovered about £150,000.

Although, in numerical terms, Amec recovered substantially less than it had claimed, the reason was almost entirely because the method of valuing the lost hours claim was decided to be by way of a valuation rather than by way of direct costs, albeit that the valuation exercise, involving fixing an hourly rate for each of the lost hours, took into account the direct costs involved. Amec succeeded on every other issue and, as I have already indicated, the variations claims were settled for a substantial sum, less only the disruption element which was transferred to the lost hours claims. Effectively, I allowed a rate per hour between £10 and £20 less than those claimed.

> Stork's minimal recovery shows how comprehensive Amec's success really was. Since the counterclaims, as I have already decided, where merely the obverse of, and set up as pure defences to, Amec's claims and since so little was recovered by Stork, I regard Stork's limited success as disentitling it to any costs on the counterclaim and as not requiring any reduction in costs otherwise due to Amec."

HHJ Thornton had some important observations on Amec's 'Analytical Methods' used in their claims,

> **"Impacted Programmes**
> Although the impacted programmes were hardly relied on at the trial, the expenditure incurred on their preparation was largely productive. The costs of preparing evidence associated with the impacted programmes were a relatively inconsequential add on to the costs of preparing base programmes, which were a necessity, and as built programmes, which were highly desirable. thus, the work and costs involved in the pure impacted exercise could not have been substantial.
>
> However, of greater importance than the actual quantum of costs is the consideration that Amec had to devise from scratch a series of exercises necessary to show that Stork's complete denial of Amec's disruption case and its refusal to provide any positive input into the question of how that case should be examined, compiled and presented. The extent of the disruption and of the primary evidence available to found the proof of that disruption were unprecedented in their scale and it is not surprising that some of the analytical methods adopted were unsuccessful. The attempt to use impacted programmes was not unreasonable and the use of expert evidence, including that from Dr Cree, was a reasonable adjunct to Amec's case.
>
> No reduction for Amec's costs of preparing impacted programme evidence should be made.
>
> **Work on the Schedules**
> Stork contends that the schedules were impenetrable, largely unnecessary and in any case only evidence of potential disruption rather than of actual disruption. The experts spent many hours discussing these schedules which, at trial, were found to have little evidential value.
>
> I do not accept Stork's strictures about these schedules. It was necessary for Amec to undertake a formidable exercise in analysing the primary facts. The exercise had both a positive and a negative utility. In positive terms, it enabled the extent, scope and size of the disruption to be demonstrated, largely by means of the experts' agreements which were founded on them. Negatively, the exercise highlighted how improbable was Stork's positive case once this emerged. It is true that the method of presenting the schedules without a clearly drafted document with them which guided the uninitiated reader through them was extremely unhelpful. However, in the

context of this case, that is a trivial objection and was overcome since the experts, in their many meetings, explained the content and scope of these schedules to each other and reached a series of extensive and helpful agreements which shaped and, ultimately, led to the overall result of the case.

Academic Research
This ground of objection, based on the formulaic methods of assessing disruption, fails for the same reasons as the impact programme objection failed. Furthermore, Stork failed to take the obvious step open to it at an early stage, namely of suggesting that the parties jointly appointed Professor Horner as an expert, whose work largely supported the formulaic methods relied on. Had Professor Horner been appointed jointly and asked to report to the parties jointly at an early stage, the method would have been dropped before many of the costs complained about by Stork were incurred.

Amec's incurring of costs under this head was reasonable and no deduction should be made for these costs.

Stork's Costs
Stork adduced detailed evidence from a costs draftsman as to the costs it had incurred in dealing with the perceived change of case and in dealing with the abandoned parts of Amec's case. This exercise was largely valueless since the costs draftsman did not describe in factual terms what the wasted work was or how he assessed the times and costs attributed to it. However, more generally, Stork is entitled to none of these costs. Since Amec was reasonable in putting forward a variety of means of analysing the evidence, even those that were unsuccessful in their aims, Stork cannot reasonable recover its costs, particularly in the absence of having put forward reasons at an early stage why those methods should not be developed, or as to how the disruption should be analysed or as to how the methods of analysis with which it had doubts could be developed jointly using one expert team reporting objectively to both parties.

Stork is entitled to none of its claimed costs."

2.3.2 Commentary

In the judgment, HHJ Thornton gives many important observations regarding the presentation and supporting information necessary for disputes concerning disruption and acceleration. Such as, Paragraph 25, which details *"The Problems Confronting the Claims Team"*.

In paragraph 45, HHJ Thornton detailed the 'Analytical Methods' used by Amec in their claim presentations.

2.3.3 Johnson Control Systems Ltd. v Techni-Track Europa Ltd.

2.3.3.1 The Facts

The first claim is by the Claimant (Johnson) for repayment of sums paid to the Defendant, Techni-Track Europa Ltd ("TTEL") under three subcontracts for the installation of electrical equipment. Glaxo Group Research was constructing a site, which comprised six buildings on 78 acres at Stevenage. The main contractor was L.M.K. Joint Venture. Johnson was subcontracted to install and build controls, controls wiring and carrier systems in three buildings known as Biology, Chemistry and C.R.S.F. There were three separate subcontracts with identical terms for each of the buildings. The relevant purchase orders were dated January of 1993.

There is a counterclaim under the subcontracts for unpaid valuations certified to be due to TTEL, for work undertaken between the date of the last valuation and the date when TTEL left the site, and for variations, delay, disruption, loss and expense and the cost of additional work.

Johnson's claim is for the repayment of monies it alleges were overpaid and the costs of completing the subcontract works. It alleges that TTEL was in repudiatory breach of contract by removing its employees from site on 9 February 1994.

Johnsons are a US firm that conducts business all over the world. They enjoy a special relationship with Glaxo insofar as they are a preferred contractor and supplier. At Stevenage they were a subcontractor to L.M.K. Joint Venture. TTEL were one of approximately 20 sub-subcontractors in relation to the three buildings.

It was made clear from the outset that the sub-subcontract between Johnsons and TTEL incorporated the terms of the subcontract between Johnson and L.M.K. Joint Venture. The scheme of payment was pay when paid.

The counterclaim is divided into three parts:

- unpaid valuations to 18^h January 1994
- unpaid work after 18 January 1994. and
- sums due in respect of delay, disruption, loss and expense incurred by the defendant in carrying out subcontract work in consequence of instructions or unproductive working practices instructed by the claimant.

"The Judgment

43. The counterclaim is divided into three parts. Unpaid valuations to 18th January 1994, unpaid work after 18th January 1994 and thirdly, sums due in respect of delay, disruption, loss and expense incurred by the defendant in carrying out sub-contract work in consequence of instructions or unproductive working practices instructed by the claimant.65. I observe that non-productive

overtime was in fact paid by Johnson to TTEL on occasions during the currency of the sub-sub-contract. The terms under which payment can be made for overtime working are set out in Clause 6.14:-

"No Overtime other than Casual Overtime shall be worked without the prior written approval of the Contractor. The request to work such Overtime shall be made in writing at least four (4) Working Days in advance of the planned performance of such Overtime. Additional payment for such approved Overtime will only be made on the prior written agreement of the Contractor.

Casual Overtime by the Sub-contractor or all Sub-sub-contractors shall be the Sub-contractor's responsibility and may be performed without prior notification to the Contract or within the Working Day solely at the Sub-contractors expense."

"THE LOSS AND EXPENSE CLAIM

66. It is clear from, the evidence of Mr Bull, Mr Lewis and Mr Ahluwalia TTEL's Contracts Manager that the Sub-sub-contract works were subject to some disruption and some delay. For the defendants to succeed on their counterclaim in this respect, they must establish a contractual entitlement to payment in respect of such delay. Some delay is always inevitable on a construction site when materials and equipment are supplied by various suppliers and a variety of trades are involved in installing systems as diverse as electrics and pneumatics. In installation work for highly specialised laboratories, and on the best-regulated construction sites, some disruption is inevitable, despite the finest co-ordination systems that can be put in place. Personnel are also subject to illness or accident. There are many events which may have a programming consequence and are part of the general risks and contingencies covenanted for by a sub-contractor for which he will wisely price for or agree special arrangements to forestall."

69. TTEL's case is refined insofar as they identify specific causes of delay leading to loss and expense in terms of lost productivity in the document sent under cover of a letter dated 6th January 1994 by Mr Bull to Mr John Waddelove of Johnson.

70. The introduction to that document is in the following terms:-

"We have undertaken the controls electrical installation as sub contractor to Johnson Control on the CRSF building, the Glaxo site in Stevenage. We have, during the contract, experienced delays, disruption and uneconomical working which has resulted in us incurring substantial additional costs, which we are seeking to recover.

The purpose of this document is to provide further details and substantiation in respect of our additional costs in order that they may be accessed and agreed in a fast and efficient manner. In order to maintain the site progress, it is essential that we are able to quickly resolve the claim situation to enable additional funds to become available against the project."

71. Later in the section 'delays and disruption' the following narrative appears:
Delays and Disruptions
We believe that during the course of the contract we have experienced severe delays and disruption to our works, which could not have been foreseen or envisaged at the tender stage. We have progressed diligently with our first and second fix activities and have suffered in terms of productivity as a direct result of site conditions. We are currently experiencing severe delays of our final terminations as a direct result of our of sequence work in respect of late delivery of control panels and controllers and the non-completion of works by other trades, including trunking, duct work and fume cupboards.

We have provided a comprehensive schedule of completion of our electrical installation and have detailed areas that are being delayed although work is outstanding by others causing disruption and associated uneconomical working. We also highlighted WERE (sic) possible cover up dates in respect of our first fix and wiring works which we believe clearly demonstrate that we have been delayed in completing our works for several weeks. It should be noted that the schedules provided are not fully complete and that additional work is being added on a number on an ongoing basis and will be updated in due course.

We believe that the procedures associated with the cover up notices were not highlighted within the specification at tender stage and we are still of the opinion that they are not clearly identified in the specification now. The associated administration disruption and substantial loss of productivity associated with these procedures are significant and we feel entitled to recover costs in respect of lost productivity.

We believe we have demonstrated from the enclosed schedules that the majority of our work has revolved return visits, out of sequence working and a significant degree of disruption in order for us to complete out works., We feel justified in claiming the lost productivity associated with the out of sequence working and the associated disruption. We have highlighted the cost in the next section associated with lost production man hours."

"LATE DELIVERY OF CONTROL PANELS
72. Johnson disclosed a schedule showing delivery of panels. I am satisfied that it is accurate.
It confirms that some panels were late delivered.
....
74. No excessive margin of re-visit over and above contractual scope has been proved in this case. Furthermore, as to loss of productivity there is no evidence that tradesman were standing around with nothing to do. The evidence of those on site, employed by TTEL Mr Lewis and Mr Hassell is to the contrary."

"CEILING GRID ERECTION
78. In order to fit conduits, controllers and other equipment in the void, TTEL's tradesmen may have had to wait for ceiling fixers to remove tiles to enable them to have any access. Fitted grids would cause marginally more difficulty in any event in getting free access to the roof void. TTEL however, have not identified

building by building, room by room where the difficulty was, who caused that difficulty and what impact it had day by day upon their productivity. Were operatives stood down because there was no work for them to do? Did they have to work extra time? There was no evidence led by TTEL.

79. TTEL have been unable to demonstrate the degree of any delay, or who caused it, or whether it was a TTEL's failure to follow a programme, or Johnsons failure or another trades failure, or whether any was such a delay was within the original scope of works."

"COVER UP NOTICES

82. There was no attempt by TTEL to particularise its case as to disruption causing delay or loss of productivity in consequence of particular notices being late, or batches of notices, in relation to particular areas and demonstrating the consequence of any lateness on its programme.
83. TTEL failed to demonstrate that the lateness of cover up notices caused any particular delay or loss of productivity."

"THE EXPERT EVIDENCE

92. As to the valuation of TTEL's claim for loss and expense, disruption and delay, he endorsed TTEL's approach as a valid one. He referred to the information sent to Johnsons under cover of the letter of 6th January 1994 as being simply a record of the progress of installation to the dates in mid-December of 1993, together with calculations of costs. He confirmed that these costs were based on a comparison of the costs of the quantity of man-hours recovered through applications made by TTEL to Johnson against the costs of the actual man-hours used on site, enabling a comment on the levels of productivity to be made.

He said in his report:
"The calculation of the actual man-hours was based both on the record and time spent on various buildings, as recorded on TTEL's time-sheets and backed up by the daily records of labour on each element of the project as provided to LMK and via JCS by TTL. The calculation of recovered hours is deduced from the gross values of agreed applications made by TTL, having made allowance for materials content of the application, and using, the tendered labour rate, the hours are calculated.

TTL then calculate the apparent losses on labour committed to the contract by using the lost man-hours calculation, price and the rate included within the tenderers/sub-contract orders for the projects. In addition they projected the probable losses in terms of man hours to complete the contracts, in order to indicate to JCS the possible total costs:

In addition to the calculation for loss of productivity/man-hours, TTL included some additional costs on the chemistry project due to changes in drawings from tender stage when compared to actual construction drawings issued to them. Arguably these additional costs would not necessarily form part of the loss and expense claim, but would be recovered as a variation item to the various projects. However, it must be remembered the impact of such variation may lead to prolongation of the contract, to a claim by TTL for such prolongation.

These methods of calculating loss productivity are often used in the mechanical services section of the Construction Industry for the calculation of additional costs due to delays etc. Had the Contract not been repudiated, I would assume that JCS and TTL would have negotiated some form of settlement around the information provided by TTL, possibly with JCS taking such information forward as part of their claim for delay, destruction etc, against their employer ELMK:

I see no flaw in the method and calculation used by TTL, but would note that ultimately a settlement figure for such claim for delay, destruction etc.., would probably be lower than the figures included by TTL in their documentation......

I accept that that may be a valid starting point to value a delay and disruption claim where causative delay and disruption is demonstrated. Mr Quigley was unable to establish the responsibility and reasons for the delays said to lead to disruptive or unproductive working advised by TTEL. He sampled a number of pages of the claim document, namely every ten pages, because he was unable to carry out any detailed analysis of the individual items of the claim which might indicate the state or extent of the work completed on each item in order to establish the party with whom the responsibility for the apparent delay lies.

94. Mr Quigley in this sampling exercise allocated the party to whom he believed the responsibility or delay or non-completion of the work lay. The results show that as a percentage of the total items sampled, TTEL appear responsible for 1% of the number, unknown 47% and JSL 52%. He concluded that given that the document was prepared by TTEL the low level of their apparent responsibility is of no surprise. He then suggested that the unknown portion would have to be split between either TTEL and Johnson, as from their point of view they were the only parties to the sub-contract between them. He entered the caveat that the percentage split indicated made no allowance for the weighting of the responsibility for delay based on the importance of the item concerned.

95. An assessment of probabilities is a matter of evidence. The evidential basis for such a conclusion has not been established. As I have observed above in relation to each aspect of this claim for loss and expense TTEL has failed to demonstrate any occasion of unforeseen delay or event of disruption causatively leading to recoverable loss and expense under contract. I reject Mr Quigley's speculative and theoretical approach to assessing responsibility under the contract for the general delay complained of. It is arbitrary and ignores criticality."

"THE TERMINATION OF THE CONTRACT

108 Johnson have never made any criticism as to the quality of TTEL's work. That is of significance in the light of criticisms made by Johnson of the fact that many of the personnel used by TTL on site were self-employed and provided through an agency. It was said by Mr Ahuwalia, TTEL's contract manager that agency labour was not motivated, and that there was a rapid turnover and new entrants had to familiarise themselves with the job and the high standards required. This may have led to reduced productivity."

2.3.3.2 Commentary

In the judgment, HHJ Wilcox gives some important observations regarding the presentation and supporting information necessary for disputes concerning disruption and loss of productivity.

With regard to 'disruption' and 'loss of productivity', HHJ Wilcox commended the Defendant, TTEL, for informing Johnson Controls during the project that they were experiencing disruption to their works and giving clear examples(see paragraphs 69, 70 and 71).

However, he was critical of TTEL in respect of their complaint with 'Cover-Up Notices' insofar that they did not particularise the causes and effect of the disruption or loss of productivity caused; stating in paragraphs 78, 79, 82 and 83:

---

78. "In order to fit conduits, controllers and other equipment in the void, TTEL's tradesmen may have had to wait for ceiling fixers to remove tiles to enable them to have any access. Fitted grids would cause marginally more difficulty in any event in getting free access to the roof void. TTEL however, have not identified building by building, room by room where the difficulty was, who caused that difficulty and what impact it had day by day upon their productivity. Were operatives stood down because there was no work for them to do? Did they have to work extra time? There was no evidence led by TTEL.
79. TTEL have been unable to demonstrate the degree of any delay, or who caused it, or whether it was a TTEL's failure to follow a programme, or Johnsons failure or another trades failure, or whether any was such a delay was within the original scope of works."
82. There was no attempt by TTEL to particularise its case as to disruption causing delay or loss of productivity in consequence of particular notices being late, or batches of notices, in relation to particular areas and demonstrating the consequence of any lateness on its programme.
83. TTEL failed to demonstrate that the lateness of cover up notices caused any particular delay or loss of productivity.

This is a valuable reminder from the Court to those preparing and submitting disruption claims to provide cause and effect particulars with such claims.

HHJ Wilcox had some unflattering comments on the expert evidence in respect of the sampling exercise of the claim document to establish responsibility and the allocation of loss and expense of disruption (see paragraphs 93 and 94 of the Judgment above).

His final comment on the matter of 'disruption' in this trial is given in paragraph 95 of the Judgment as follows:

> "An assessment of probabilities is a matter of evidence. The evidential basis for such a conclusion has not been established. As I have observed above in relation to each aspect of this claim for loss and expense TTEL has failed to demonstrate any occasion of unforeseen delay or event of disruption causatively leading to recoverable loss and expense under contract. I reject Mr xxx's speculative and theoretical approach to assessing responsibility under the contract for the general delay complained of. It is arbitrary and ignores criticality."

Again, here is a valuable reminder from the Court to those preparing and submitting disruption claims to provide cause and effect particulars, with actual events and circumstances, if at all possible, with such claims.

2.3.4 Cleveland Bridge UK Ltd v Severfield-Rowan Structures Ltd

2.3.4.1 The Facts

This case relates to steelwork for the Shard. Cleveland Bridge UK Ltd ('CBUK') was the fabricator and supplier of steelwork for the first nine levels of steelwork, engaged by Severfield-Rowen Structures Ltd ('SRS'). The primary issues between the parties relate to alleged delays on the part of CBUK in the provision of such steelwork and to the impact of any such delays on the progress of SRS in erecting the next 31 levels of steelwork.

SRS was carrying out major projects, amongst many others, in parallel with the Shard. CBUK was itself a substantial steelwork contractor, with extensive experience in the fabrication and erection of steel. It was also extremely busy on other substantial projects throughout the country.

There were to be five relevant tower cranes: TC1 was located in or on the central concrete core, TC2 in Phase 1 at the western edge, TC3 in what was called the 'Backpack' area, TC4 at the northern end of Phase 2 and TC5 at the western edge of Phase 3. Under its contract with Mace, SRS was entitled to 80% of the working hour usage of TC2, TC4 and TC5 and 60% of TC3.

Logistics at the site were restricted. There was a one-way system, with delivery lorries coming in from the east side and delivering their materials to various pick-up points on the south side from which the cranes could offload materials. There was little or no storage space for steel, which needed to be offloaded by one of the adjacent cranes and then taken directly or via another crane to its location in the right phase at the right level for erection. The steel would be erected by steel erectors often standing in a type of cherry picker (a Static Elevating Working Platform –SEWP), who would receive each piece of steel as it was manoeuvred into position by the crane and locate and fix it into position. Once sufficient of a floor or level was completed, the decking would go onto the floor beams, the SEWP could be relocated onto it and steel work above it could be started.

So far as the logistics of steel fabrication in this case are concerned, full design information needed to be provided to CBUK which had to produce detailed fabrication drawings for approval; the steel then had to be fabricated by CBUK to the right size and shape, painted with a primer and delivered from the CBUK works at Dalton to SRS' premises in Thirsk. There SRS would 'shake out' the delivered steel into the sequential erection Lots, send the steel through its own paint shop for painting and curing as necessary and then store it or load it onto a truck for delivery to site.

Mace Limited ('Mace') was the main contractor employed to build the Shard. Initially, Mace retained CBUK in or about 2008 pursuant to a letter of intent to secure steel and to carry out various initial steel works. In early 2009, however, Mace invited several steelwork contractors, including CBUK and SRS, to tender for the steelwork overall. Tenders were submitted and there clearly was liaison between Mace on the one hand and SRS and CBUK on the other. In March 2009, CBUK and SRS were separately invited by Mace to provide a revised tender on the basis of revised tender documentation dated 12 March 2009. Both were referred to the proposed Contract Programme, then Revision 8, and to the requirements and general location of where the various cranes were to go. They were also referred to the allocation of crane time to the steelwork subcontractor, which included the allocation during the normal working hours for the Main Tower which were Monday to Friday 8 am to 6 pm and Saturday 8 am to 1 pm.

SRS was successful and CBUK was not. On 30 April 2009 Mace instructed SRS to proceed initially with what were called the 'Early Works' which included 'material transfer', 'raw steel production' and the commencement of the fabrication of ground to Level 9 steelwork, pending a final negotiation of the subcontract. It was a 'letter of intent' and envisaged a start date on site of 23 November 2009. The 'material transfer' related to the transfer of steel from CBUK which it had secured pursuant to its earlier letter of intent.

On 1 May 2009, CBUK tendered to SRS for the fabrication of steelwork up to and including Level 9. It quoted for 3826 tonnes of steelwork at £617 a tonne, together with drawing office work and blasting and priming all steelwork. It excluded 'shelf angles and decking'. It anticipated completion by the end of November 2009. It identified that 'detailed contract specific terms and conditions [are] to be agreed'. The tender had been based on 250 specific drawings.

By e-mail dated 7 May 2009, SRS sent to CBUK a letter of intent for the steelwork fabrication up to Level 9. The e-mail anticipated that the 'formal order will be forwarded during the next few days'. The scope was to include all fabrication and the basic steelwork necessary to be considered as having been procured by CBUK under its previous arrangement with Mace; other 'black steel' would be provided 'free issue' by SRS. Rates were identified, as quoted for.

On 9 May 2009, Mr Robinson of CBUK sent a fabrication programme which showed fabrication commencing towards the end of June 2009 for the lowest levels completing Levels 8 and 9 by 25 November 2009.

On 28 May 2009, Mr Robinson sent to Mr Willis an 'off-site programme which now identifies the drop dead dates we require information by'; this was FAB 002 Rev 1. This showed all information for Levels 0–4, 5, 6–7 and 8–9 by July 5, 12, 26 and 10 respectively with fabrication commencing for the lower levels on 22 June and all to be completed by 25 November 2009. This was reviewed by SRS over the following few days.

2.3.4.2 The Judgment

HHJ Akenhead's Judgment contains his observations on disruption and acceleration, together with important observations on mitigation. The paragraphs containing these are in the section on Disruption.

For his quantification of the disruption caused to SRS's works, their Expert was unable to use either of the two most recognised methods, the 'measured mile' or 'earned value' approaches.

As stated in paragraph 155 of the judgment:

> "He did not have appropriately detailed timesheets or daily record sheets to enable standing or non-productive time to be identified. He could not tell from the documentation whether the steelwork deliveries to site resulted in downtime and lost erection hours. He could not find a resource schedule or resource planning documentation within SRS or indeed Steelcraft to see what was allowed for. Finally, he says unequivocally in his first report at Paragraph 10.12.23 that he had not located detailed records for Steelcraft labour that would facilitate any, or any meaningful disruption analysis. He accepted in cross-examination that it was at least logically possible that the reason for the increased number of shifts was inadequate planning, inefficient use of cranes or even that Steelcraft was not very good at its job. There is limited reliable contemporaneous evidence that the so-called planned resource of 630 shifts was the planned resource; it was supported by Mr Willis but without reference to any contemporaneous document and it was contradicted by the original pleading which identified a higher planned resource."

> 119. SRS' counterclaim falls broadly into two categories, costs and losses associated with the delay in deliveries by CBUK and defects; it now accepts the figures put forward by Mr Gurnham. ....
> 141. Essentially, I find it difficult on the evidence to conclude that delay caused by any of the crane downtime was, on the balance of probabilities, attributable to CBUK's late delivery of steel. Some of it may have been so attributable but it has simply not been proven. Counsel for CBUK highlighted in cross-examination a

number of other factors which could have caused delay and further disruption. For instance, there was reference within Mace documentation and indeed within SRS' disclosure to variations issued after April 2010 at least possibly delaying progress. In circumstances where (even for understandable reasons), there has been no critical path analysis and in reality no detailed and sequential evidence explaining what happened as the steelwork went up to Level 40, it would be wrong for the Court to infer that in effect the large bulk of the factors which in truth led to the recovery acceleration programme not being achieved and indeed to further delay can be causatively attributed to CBUK's failure to deliver steel by the last week in February 2010. In practice, SRS has not proved that these other factors such as crane downtime would not have occurred in any event, irrespective of any fault on the part of CBUK. SRS has not proved that the need to remedy defectively fabricated CBUK steelwork caused any overall delay or even that it disrupted the accelerative effort embarked upon by SRS.

Disruption
152. SRS' pleaded claim is in the sum of £94,500.50 and is predicated upon the basis of "the increased labour costs incurred by SRS as a result of inefficiencies caused by intermittency of working because of the delayed delivery of steel" (Paragraph 36 of Further Information). SRS goes on in that it can not "say which individual events caused which individual periods of disruption" but that the "delayed and piecemeal delivery by CBUK to SRS at Thirsk of all pieces of steel to each particular phase to SRS resulted in disruption to progress and inefficient workings of the steelwork subcontractor and the metal deck installation subcontractor". The pleaded quantum is predicated upon "Steelcraft's planned erection resource for the steelwork floors 5 to 9 [being] 786.67 man shifts over the period from 8 February 2010 to 18 March 2010" and "the actual resource level being 1137.65 man shifts", with the resulting balance of 350.98 man shifts at a rate of £219.43 per man shift producing the largest element, £77,015.54. Apart from that claim, there are claims for plant (estimated at 5% of labour costs - £3,850.78), and for Steelcraft overheads and profit (£8,806.35 and £4,827.73 respectively). This pleading was amended, belatedly, on 1 November 2012 to increase the disruption claims to £225,055.43 based on there being 1145.83 man shifts as against 630 shifts planned by SRS.
153. Essentially, Mr Gurnham has assessed this claim as follows: ...
154. Mr Gurnham was clearly in a quandary about this claim. He accepted openly in his report that he had not been able to use any of the normal approaches to evaluating disruption. For instance he had not been able to use the "measured mile" or "earned value" approaches, these being commonly used to assess disruption loss. He did not have appropriately detailed timesheets or daily record sheets to enable standing or non-productive time to be identified. He

could not tell from the documentation whether the steelwork deliveries to site resulted in downtime and lost erection hours. He could not find a resource schedule or resource planning documentation within SRS or indeed Steelcraft to see what was allowed for. Finally, he says unequivocally in his first report at Paragraph 10.12.23 that he had not located detailed records for Steelcraft labour that would facilitate any, or any meaningful disruption analysis. He accepted in cross-examination that it was at least logically possible that the reason for the increased number of shifts was inadequate planning, inefficient use of cranes or even that Steelcraft was not very good at its job. There is limited reliable contemporaneous evidence that the so-called planned resource of 630 shifts was the planned resource; it was supported by Mr Willis but without reference to any contemporaneous document and it was contradicted by the original pleading which identified a higher planned resource.

155. Against all these difficulties, it is inevitable that there must have been some disruption to erection progress in the February to April period. Deliveries by CBUK were late and in some respects out of sequence. This led to delays and piecemeal deliveries to site. There must have been disruption in the sense of reduced productivity. It is beyond doubt that Levels 5 to 9 should have been erected within the weeks commencing 8 February and 15 March 2010. I accept also that there is reliable evidence that the planned resource was related to using 20 men per week on 5½ days per week. There were thus going to be 110 man shifts per week. Within the erection programme, the steelwork up to and including Level 9 should have been erected and capable of erection utilising 630 man shifts identified by Mr Willis; this has been verified by Mr Gurnham. However, 808.50 man shifts were deployed up to 18 April 2010. I disregard the period after this because the further delays were caused by factors for which CBUK are, in the light of my other findings, not responsible.

156. It follows that the disruption attributable to CBUK can not have exceeded the difference between 808.50 and 630 man shifts, that is 178.50 man shifts. It would be wrong however to base any assessment of the probable disruption cost on this balance by reason of the absence of detailed evidence attributing a lack of production or productivity to CBUK's breaches of contract. What the Court can and should do in circumstances where it is satisfied on a balance of probabilities that some (more than de minimis) disruption must have occurred as a result of CBUK's breaches is to make a reasoned assessment albeit based on the minimum probably so attributable. I am satisfied that a minimum of 40 man shifts' worth of time must have been wasted. One has to bear in mind that not only were there 20 or so Steelcraft men working the steel erection during this period but also there were deployed other steel erectors at night and at weekends; therefore, notwithstanding that there were substantially more man shifts worked in this period even than the 808.50 attributed to Steelcraft alone, progress nonetheless remained painfully slow, the whole operation ended up

taking a minimum of 42 days longer than it should have done by reason of CBUK's breaches. Mr Barry gave evidence that the erection programme was achievable and I accept that evidence. There was some limited complaint by Mace about SRS' activities and there was some tangential evidence (the Leanconsult report) about some incompetence but it is inconceivable that there would not have been more complaint from Mace which was closely supervising the work if SRS steel erectors were "dragging their feet" unduly. 40 man shifts represents the average of about half of one week's worth of man shifts during this period.

157. Accordingly, I find that 40 man shifts' worth of disruption cost was incurred the rate established by Mr Gurnham (which I accept), namely £217.84, and that £8,713.60 consequentially represents the disruption labour cost attributable to CBUK's breaches of contract. As for the addition of 21.63%, elements of this at least are more obviously recoverable with regard to ordinary labour to which this disruption claim relates. Mr Gurnham analysed this in his Appendix 3.4 to his first report and I broadly accept what he says there, based as it is on an analysis of accounting information from Steelcraft. ....

158. It will follow that SRS has established an entitlement to £10,256 (£8,713.60 + 17.71%).

SMD Disruption

159. SRS counterclaimed £372,951.76 for disruption caused to its metal decking subcontractor, SMD, and the additional £40,500 for haulage by SMD. This has been reduced in practice to the sum of £73,529.63, based on Mr Gurnham's evaluation. Although the higher sums represent what was apparently claimed at least at one stage by SMD against SRS, there is no doubt that a final account and claims settlement was reached between SMD and SRS in the sum of £1,591,157.95, as against the sum of £1,968,589.56 claimed by SMD. This settlement is complicated by the fact that SMD was engaged to carry out decking not only to the main tower but also to the "backpack" part of the Shard (with which these proceedings are not concerned) and there were claims in relation to the backpack.

Summary of Counterclaim Entitlement

205. Consequential on the findings set out above, the total sums due to SRS by way of damages are as follows: …

Conclusion and Decision

206. On the basis of the findings above, CBUK is entitled to £928,472.55 plus VAT from SRS by way of sums due under the Sub-contract against which SRS is entitled to by way of damages in the sum of £824,478.49 for delay and defects. In addition there are as yet unquantified findings of liability against CBUK for various NCRs. The VAT due to CBUK will only be due on any net balance after taking into account all damages due to SRS.

2.3.4.3 Acceleration

Some time before 28 May 2010, the London Borough of Southwark had approved night-time working in relation to steelwork from Mondays to Saturdays. SRS had done night-time working in March but had withheld full night-time working until it had been formally approved. Steelcraft subcontracted the night time and extra weekend working to two companies, MCS and K-Len, although the former had been retained in March 2010. Much of the acceleration claim relates to their costs.

> **The Counterclaim**
> 119. SRS' counterclaim falls broadly into two categories, costs and losses associated with the delay in deliveries by CBUK and defects; it now accepts the figures put forward by Mr Gurnham. I will first consider the delay costs which can be broken down, in the figures finally put forward by Mr Gurnham, as follows: ….
> 123. The programming experts, Mr Holloway and Mr Barry, agreed in their first Joint Statement that the "commencement of steel erection to Levels 8/9 (i.e. phases 801, 802 & 803) will represent the key milestones against which any delay to SRS's works caused by a delay in the supply of CBUK steel should be assessed" and that the "commencement of steel erection to Levels 10/11 (i.e. phases 101, 102 & 103) will represent the key milestones against which (a) the critical path of SRS's works can be assessed and (b) any delay to SRS's works caused by fabrication errors should be assessed." Although the two experts disagreed as to which Programme should be used against which to measure delay, they did agree that "at the most appropriate method for delay analysis is As-Planned versus As-Built Windows Analysis" as "this method allows us to identify the actual critical path" and it "is the longest path that the actual delays to completion will be located."
> 124. Having agreed this, Mr Holloway then abandoned this sensible agreement, which was unfortunate and indeed unhelpful. Mr Barry, however, broadly stuck to the agreement and analysed the delay at least up until the commencement of Phase 103 at Levels 10/11. Neither expert either did or was readily able to analyse the delays to the remainder of the steelwork erection up to Level 40. This is because the history and the facts were all very confused and confusing because SRS implemented a substantial acceleration programme to try to make up the lost time as at April 2010 and because not only was that acceleration ultimately unsuccessful in that no time was made up but also additional delay occurred.
>
> The history of delay becomes somewhat confused after mid-May 2010. Although SRS instituted extensive accelerative measures in terms of night-time, weekend and extended hours working which, all things being equal, should have had the effect of recovering all or some of the 56 days delay, SRS in

fact finished even later. Mr Barry identifies that Phase 401 finished 67 days late, Phase 402 54 days late and Phase 403 69 days late. Not only was the acceleration in the result wholly unsuccessful, matters deteriorated. He has provided at Paragraph 207 of his report a helpful analysis which demonstrates however that on 14 of the phases steelwork was completed in less than or the same as the planned durations. However on all the others work took longer than the planned duration; in some instances, there were very substantial delays. For example Phase 342, planned to take nine days, took 34 days, a delay of 25 days; Phases 362 and 381 were each delayed by 21 days over the planned durations. No explanation or evidence was provided to explain why these extended durations happened or whether they would have happened in any event irrespective of CBUK's original default.

138. Mr Barry then seeks to identify what might reasonably have been allowed within any sensible programme for downtime with a view to ascertaining the excess amount of crane downtime occurring during the erection of Levels 10 to 40. The balance he then attributes to the original CBUK delay on the basis that CBUK's delays have been such that SRS has been pushed into a period in which by reason of its acceleration measures (in particular working at nights and weekends) it has been subjected to a higher than reasonably allowable amount of crane downtime; this is attributed to CBUK because, as suggested by various factual witnesses for SRS, particularly Mr Willis, the acceleration was primarily required by reason of CBUK's culpable delay, the accelerative measures resulting in the nights and weekends not being available for most of the climbing, other and miscellaneous reasons and the programme was being pushed into a period (particularly September through to December 2010) when cranes were more prone to downtime due to weather than would otherwise be the case. To be fair to Mr Barry, under cross-examination and indeed some questioning from the Bench, he did accept that some potential reservations can be made to his thesis on this.

139. For instance, there were a significant number of phases in the July to September 2010 period when significant time was lost against the recovery programme, notwithstanding favourable crane availability. Although Mr Barry identified some (documentary) evidence which suggested possible causes of the delay during this period, he was not able to be particularly positive about this stage of the steel erection as to why it was being delayed. Additionally, when it was pointed out to him that a number of the events which caused crane downtime actually took successive days to address, he had to accept (properly) that, irrespective of whether there was a need for acceleration, extra delays would in any event have occurred.

141. Essentially, I find it difficult on the evidence to conclude that delay caused by any of the crane downtime was, on the balance of probabilities, attributable to

CBUK's late delivery of steel. Some of it may have been so attributable but it has simply not been proven. Counsel for CBUK highlighted in cross-examination a number of other factors which could have caused delay and further disruption. For instance, there was reference within Mace documentation and indeed within SRS' disclosure to variations issued after April 2010 at least possibly delaying progress. In circumstances where (even for understandable reasons), there has been no critical path analysis and in reality no detailed and sequential evidence explaining what happened as the steelwork went up to Level 40, it would be wrong for the Court to infer that in effect the large bulk of the factors which in truth led to the recovery acceleration programme not being achieved and indeed to further delay can be causatively attributed to CBUK's failure to deliver steel by the last week in February 2010. In practice, SRS has not proved that these other factors such as crane downtime would not have occurred in any event, irrespective of any fault on the part of CBUK. SRS has not proved that the need to remedy defectively fabricated CBUK steelwork caused any overall delay or even that it disrupted the accelerative effort embarked upon by SRS.

143. It follows that the reasonable costs incurred by SRS in implementing acceleration measures to overcome the 42 days delay caused to the steel erection programme by CBUK's failures to deliver steel in accordance with the December Programme should be recoverable. The fact that the acceleration was not ultimately successful due to factors which SRS can not establish are attributable to those failures can not and should not prevent or limit recovery of such costs. However, there are necessarily on the facts of this case some qualifications:

   (a) One needs to take account of the fact that the recovery programme was adopted to overcome not only the 42 days delay caused by CBUK's breaches but also the further 14 days delay which is not attributable to those breaches. Both parties' experts and indeed Counsel accepted that the appropriate way to proceed in this context was to determine the overall figure attributable to acceleration and then discount it, in the context of these findings, by 25%.

   (b) There must have come a time between April and December 2010 when the acceleration measures (extended, night-time and weekend working), which continued throughout this period, were not in fact being deployed to overcome the 42 days delay caused by CBUK's breaches but were being deployed to prevent the delay becoming worse for other reasons not attributable to CBUK. There has been no direct evidence as such which identifies precisely when this "tipping point" was actually reached. However, in principle the costs of the acceleration after this point can not be said to be attributable to CBUK's breaches; rather, they were attributable to whatever other factors, such as crane downtime or Mace instructed variations, were actually operating to prevent the programme being recovered and in fact producing further delays.

SRS Acceleration

161. I refer to my findings above in relation to acceleration. It therefore falls to the Court to decide on the evidence what were the actual and the reasonable costs incurred by SRS in accelerating the work in what turned out to be an unsuccessful attempt to recover the delays which had occurred. Essentially, what happened was that from late February 2010, Steelcraft and its sub-sub-contractors began to work double shifts up to 11 pm with additional time over the weekend, this being implemented to seek to recover and mitigate those delays which had already been caused by the late deliveries of steel by CBUK; further night-time working began in or about March although it appears that at least initially proper consent had not been obtained from Southwark Council. There were noise trials carried out in April 2010 going into May and approval from the Council was received later in May. The recovery programme was initially based only on the extended hours (7 am to 8 am and 6 pm to 11 pm during weekdays and 7 am to 8 am and 1 pm to 4 pm on Saturdays). It is clear however that full night-time and weekend shifts were being worked over and above this recovery programme. It must be understood that working on a site such as this at night is and was necessarily less productive primarily because there is a need to keep noise down and steel erection can be a very noisy business; there were restrictions on using hammers at night. This meant that SRS could only fix steel members into position often only pinning them together with one bolt as a temporary measure until the connections were finally and fully fixed by the wrapping up team the following day. Remedial works could not be done at night if they required hammering.

162. To carry out the acceleration steel erection, Steelcraft, with the knowledge and consent of SRS, sub-sub-contracted the acceleration work to MCS and K-Len and, indeed, much of the acceleration costs claim incorporates the costs incurred in relation to and by these two sub-sub-contractors. Mr Barry said, and I accept, that the originally planned erection resources were adequate to secure compliance with the planned erection programme. In broad terms, the planned erection resources continued to be deployed but the acceleration resources were also required simply to seek to recover the critical delays which had already occurred.

163. Mr Gurnham has evaluated this claim as follows (with the pleaded figures in brackets):

Item	Amount allowed [claimed]
Day/night shift squads	£1,682,001.76 [£1,700,142.98]
Management for resources	£441,559.27 [£406,937.72]
Senior management on site	£252,417.41 [£419,549.70]
NCR Management	Nil [£135,151.98]
Total	£2,375,988.44 [£2,661,782.38]

164. Mr Gurnham allocates £233,451.57 to the period up to 13 May 2010 and the completion of Level 9 and the balance, £2,142,526.87, to the period thereafter up to Christmas 2010 when Level 40 was completed. He excludes the NCR costs which he evaluates in fact at £193,249.96 on the basis that it is not to do with acceleration.

165. However, these figures do not, understandably, take into account the findings that I have made which have the effect of not attributing to CBUK the further causes of delay which occurred from May onwards and which prevented the recovery programme from being achieved. As indicated above, there came a point when the acceleration resources were being deployed not to make up the delay which had been caused by CBUK (42 days) but to counteract the further events (be they variations, crane downtime or whatever) which were in fact causing more delay and which can not be proved to have been caused by or attributable to CBUK's breaches.

166. The problem for the Court therefore is how to assess or otherwise find what acceleration course actually incurred were caused by or attributable to the mitigation measures embarked upon as far back as March 2010 to overcome the delays caused by CBUK. It would be wrong just simply to take the total costs and artificially reduce them by some percentage because that would be simply arbitrary.

167. Mr Davis in a supplementary report (entitled "Note") put in at a late stage during the trial came up with a method of assessment which is of interest, albeit that he abandoned part of his initial calculations (Tables 1 to 3 - Transcript 24 October 2012 - page 107). His Tables 5 and 6 considered what the likely costs were in the period May to October 2010 to reflect either the recovery programme (Table 6 - with no additional nightshifts) or the actual additional dayshifts, weekend working and nightshifts actually worked in the period. In effect he says that this is a sensible way of assessing the acceleration costs attributable to any culpable CBUK delays as it reflects either the recovery programme which, all things being equal, he suggests, should or at least could have recovered the delays or the recovery programme plus the nightshifts actually worked. He compares that with what in effect would be the equivalent case as pleaded (Table 4). Table 4 produces a total sum of £881,756.05, Table 5 £731,496.42 and Table 6 £161,810.84, albeit elsewhere in his second report he adds an overhead at 10% to these figures. There is no issue between the experts in relation to the number of shifts worked by MCS and K-Len but different rates are used by the experts with Mr Davis assessing actual shift rates at a lower level than those pleaded by SRS (Table 5) and in Table 6 he uses a much lower rate because he assumes that the additional shifts should be costed by reference to the recovery programme (that is, say, only up to 11 pm at night rather than for a full night shift of 11 to 12 hours. He qualifies what he says by saying that the figures would

need to be reduced (on the basis of my findings) by 25% to reflect the 14 days delay which is not the responsibility of CBUK. His figures however do not allow for the period prior to the formal adoption of the recovery programme towards the end of May 2010.

168. So far as the rates are concerned for these acceleration shifts, it is likely that the rates were agreed between the parties with dayshift agreed at £279.20 a night shift at £356.07. It is likely that the nightshift rate applied was applicable to weekend working; I accept Mr Gurnham's evidence that these rates are reasonable and realistic and also they represent what was effectively paid or payable by Steelcraft to MCS and K-Len; the detailed breakdown was provided which Mr Gurnham has checked. Mr Davis adopts lower rates for the day shifts (£228.09) and weekend and night shift working (£297.71); he says that the rates used by SRS in this claim are "high".

169. I have formed the view that the reasonable costs of seeking to mitigate the impact of the culpable delays caused by CBUK can and should be assessed as follows:

(a) MCS and K-Len were primarily deployed by Steelcraft with SRS' consent and knowledge to seek to overcome the delays to the steelwork erection which had started to be generated by the end of February 2010. By the time that the recovery programme was put in place in May 2010, that delay was 56 days or 8 weeks of which 6 weeks was attributable to the breaches of CBUK.

(b) As from May 2010, these resources were deployed and were initially effective; indeed in June 2010 some time had been recovered. However, by September 2010 other factors had probably combined to cause further delay and as from the end of August 2010 these other factors represented the dominant reason or cause for the need to continue to retain these additional accelerative resources. Therefore as from then it cannot be said that there is any or certainly any sufficient causative link between CBUK's breaches and the need to continue to deploy the accelerative resources.

(c) Mr Davis' approach based by reference to the recovery programme is a reasonable and realistic one, albeit that his exclusion of nightshift and his rates are unrealistic. It seems to me that, although the recovery programme only extended day shifts and Saturday working, SRS was always intending to and needed to deploy what it actually deployed which was full nightshifts and full weekend working as well as to seek to recover the time which had been lost.

(d) Therefore the proper amount to allow to SRS and what it was liable to pay its sub-contractors is reflected in the following table:....

170. To this should be added Steelcraft's overhead and profit at 9% and 6% respectively, given that it was asked to provide the accelerative resources and that it

was not to blame in any way for the six weeks delay caused by CBUK which those resources were deployed to overcome. This also reflects Steelcraft's entitlement to be paid a reasonable sum for providing the accelerative resource. This produces a total of £522,659.05.

171. I now turn to the acceleration costs said to have been incurred in the period between the end of February and May 2010 and which are not encompassed by the above assessment. This was a period in which, although there is evidence of nightshift working in the period starting in the week commencing 15 March 2010 and one weekend worked (week of 8 March 2010), for which generally see Mr Gurnham's Appendix 3.10, there is some real doubt as to whether all the recorded nightshift work was actually done at night. An examination of the time sheets for March and April show that much of the time was spent installing stairs and handrails; there has been no reliable evidence to explain why the costs of dealing with this work (for which CBUK did not do the fabrication) needed to be dealt with during night shifts (assuming that that is what they were). There is something which is not wholly credible about the additional nightshifts in April 2010 because there was, unsurprisingly, given CBUK's serious delays in the delivery of Phase 803 steel, rather less to do then, such as would not obviously involve nightshifts. I am therefore not satisfied that any substantial part of the nightshift or the single weekend worked in this period is attributable to any acceleration measures which can be laid at the door of CBUK. As before at least on one claim addressed earlier in this judgment, the fact that there is a nagging feeling that some part of these extra measures might well have been attributable to the CBUK breaches is not enough to justify an allowance of damages. I bear in mind that it would have been easy for SRS to prove at least by way of representative evidence why these overtime shifts had to be worked; for instance, some evidence could have been deployed as to why MCS worked on the one weekend in the week commencing 8 March 2010: SRS did not do this. I therefore allow nothing in respect of acceleration costs in this earlier period.

172. Turning to the other elements of the acceleration claim, additional site resources and senior management, these will all fall to be reduced to exclude costs claimed as being associated with the period after August 2010 and before the date in May when the recovery programme was undoubtedly initiated, albeit with full night shift working as well as weekend working. The additional site resources claim was supported by Mr Gurnham under the following heads: …

173. Mr Gurnham has adjusted the first two items upwards and the overheads and profit has been adjusted down and up respectively, compared with the pleaded case. As I understood Counsel for CBUK, no pleading point is taken with regard to these adjustments.

174. I can and do accept that the accelerative measures needed to be and must have been supervised at a moderately or higher senior level, particularly in the context of this project being a high profile one which was going wrong. It is therefore necessary to ascertain the salary costs of these people and then to reduce them to take into account the weeks prior to May and after August to take into account the fact that it has not been proved that accelerative costs associated with CBUK's breaches are attributable thereto. That produces a total sum of £46,123.90 (made up of £15,854.89, £15,971.94, £4,899.75 and £9,397.32 in respect of the four men in question), relating to the week commencing 24 May to the week commencing the 23 August 2010. The point is made that there are no specific time sheets which allocate these men's time to the accelerative efforts and I accept that they must have been spending a substantial part of each day on the non-accelerative work which was going on during the day. Doing the best that I can, I assess that one third of their time must have been spent on servicing and supporting the accelerative efforts and therefore allow one third of the sum, namely £15,374.63.

175. I do the same exercise in respect of the nightshift where a Mr Farley was the supervisor. His salary costs are £18,603.13 for this comparable period but because the nightshift was entirely accelerative the whole of his salary costs of this period is allowable.

176. Against both these sums, £15,374.63 and £18,603.13 totalling £33,977.76, 25% should be deducted, leaving the net sum of £25,483.32, to reflect the fact that the accelerative effort was deployed to overcome 56 days delay of which only 75% was attributable to CBUK.

177. I see no reason in principle why Steelcraft should not have been entitled to a percentage mark-up on these costs to reflect its overhead and profit given that there has been no suggestion that Steelcraft itself was responsible in any way for the 42 days delay to the project caused by CBUK's breaches. Steelcraft was asked and required by SRS to institute the acceleration process and that must have been by way of additional work for which it was entitled to be paid a reasonable sum which would include overhead and profit. I accept the figures put forward by Mr Gurnham of 9% and 6% respectively.

178. Accordingly, with profit and overhead added, I find that £29,305.82 has been proved.

179. The remainder of this claim relates to specialist sub-contractors retained by Steelcraft to carry out an array of engineering, surveying and other technical services in relation to steel erection. The Court has not been shown (in one sense mercifully) the numerous invoices which support these costs which Mr Gurnham has audited and, indeed, there is no issue that the costs were incurred by Steelcraft. Mr Gurnham frankly accepts that he is unable to say one way or another whether these sub-contract resources were anything to do with

acceleration. There is no witness evidence which seeks to explain the extent to which any of them was involved in the acceleration measures. On that basis, SRS has simply not proved its case in relation to these sub-contractors.

180. Turning to the third head of claim within acceleration, the provision of senior management, there can be no doubt in my judgment that Steelcraft did deploy Mr Ashton and Mr Rennison amongst other things to manage the acceleration. In that context I accept the evidence of Mr Willis to that effect. There are two problems however relating to this claim, the first being to determine how much of their time was applied in effect to dealing with CBUK's delay and more importantly the accelerative efforts introduced to overcome those delays, a corollary of which is the extent to which they would have been deployed in any event irrespective of CBUK's delay, and secondly whether particularly in the case of Mr Ashton his time as managing director can be dealt with as anything other than as a head office overhead.

181. Turning to the third head of claim within acceleration, the provision of senior management, there can be no doubt in my judgment that Steelcraft did deploy Mr Ashton and Mr Rennison amongst other things to manage the acceleration. In that context I accept the evidence of Mr Willis to that effect. There are two problems however relating to this claim, the first being to determine how much of their time was applied in effect to dealing with CBUK's delay and more importantly the accelerative efforts introduced to overcome those delays, a corollary of which is the extent to which they would have been deployed in any event irrespective of CBUK's delay, and secondly whether particularly in the case of Mr Ashton his time as managing director can be dealt with as anything other than as a head office overhead.

182. Whilst Mr Ashton as the managing director is in one sense part of the head office overhead, there is no doubt that he spent a disproportionate amount of his time in dealing with the delays and the accelerative measures at the Shard site. He was therefore being deployed in effect as a specific and dedicated resource and therefore his time in that respect could properly be charged by Steelcraft to the project, other than as a head office overhead. The very fact that he was extensively so deployed demonstrates that. However, it was accepted in evidence that he was not full time and therefore, doing the best that I can, I assess that half of his time allocated to this project was and must have been closely related to the extensive acceleration effort that was required from May onwards. So far as Mr Rennison is concerned, he was undoubtedly allocated to this project as the site manager and would have been on the site for much of the time in any event. I have no doubt however that a significant part of his time, like Mr Ashton, was and would have been concerned with the acceleration effort and therefore it is appropriate to allocate part of his salary costs during the relevant period to that effort; but for the acceleration, he could have been

deployed elsewhere. Again, I assess that half of his time was and must have been intimately associated with the extensive acceleration effort required from May onwards.

186. It is unnecessary to consider the Mace Tower Crane cost in relation to the period up to and including the week commencing 13 May 2010 or after the week commencing 26 August 2010 because, as indicated earlier, I do not consider that CBUK's breaches can be proved to have been causative of the need to accelerate in those periods. Therefore I am concerned to consider only those costs booked against SRS for the intervening period. I disregard the first of these weeks (week commencing 20 May 2010 in which the breakdown indicates that these were shared with Byrne Bros who were concerned with other aspects of the work, given the absence of any evidence about this sharing. The resulting total is £286,884.50 which represents the maximum which could be attributed to the acceleration effort and related costs which CBUK is liable.

187. There is little or no evidence as to what the cranes were doing specifically although it is a reasonable inference (which I do draw) that much of the deployment of the cranes, operators and related staff must have been to support the accelerative measures at night, given that most of the accelerative efforts at night or at weekends and cranes were needed at night to support the accelerative efforts. There are however some anomalies. For the last three weeks in this period, the crane hours deployment goes up from much lower total hours figures for the weeks (for instance in the week commencing 5 August a total of 228 crane hours were logged) to over 300 hours with no explanation on the evidence as to why such a large increase occurred then; 324, 300 and 338 hours are logged for the weeks commencing 12, 19 and 26 August 2010. No evidence was given as to why all six cranes were deployed, for instance in those last three weeks.

188. It is therefore incumbent on the Court to make some assessment in circumstances where it must have been the case that cranes were being deployed to support the acceleration (at night) and that SRS must have incurred cost in terms of what it has accepted (properly) it is liable for to Mace. I make my assessment on the following basis:

    (a) I disallow any usage of TC1 and TC6 over this period as the expert evidence reveals that it was essentially only TC2, TC3 TC4 and TC5 which were necessarily involved in the steel erection (see Paragraph 56 of Mr Barry's report). This equates to 388 hours of crane time and to a money sum of £23,280.

    (b) In addition, I reduce the total crane hours for the weeks commencing 12, 19 and 26 August 2010 to a total of 228 hours, which allowing for the reductions made in the preceding sub-paragraph takes out an additional 48 hours at £63 per hour, nearly £3,024. The 228 hours allowance comes from the preceding two weeks and is not inconsistent with some of the other total hours in some of the preceding weeks.

(c) I exclude the hoist driver's costs because I can not see (and there is no evidence) why the hoist driver is required in circumstances where such evidence as there is suggests that the hoist was being used to enable materials other than steel to be lifted to higher parts of the building. This would remove four weeks worth of his time valued at £1,350 per week, that producing a deduction of £5,400.

(d) It is appropriate however to make a further more general reduction to reflect the absence of specific evidence as to what these cranes were actually doing and, for instance, whether opportunities were being taken to do work which was not in reality part of the accelerative effort. It is necessary, in the absence of such detailed evidence, to make a reduction which the Court can be confident adequately takes this into account. In my judgement, an adequate and safe further reduction is 50% of the balance.

(e) I therefore assess the amount attributable to CBUK as:
Cost for period: £286,884.50
Less (a): £23,280.00
(b): £3,024.00
(c) £5,400.00
Sub-total: £255,180.50
Less 50% £127,590.25
Total: £127,590.25

189. From this figure of £127,590.25, sum of 25% needs to be conducted reflect the delays which were not the responsibility of CBUK, leaving a balance of £95,692.69.

Summary of Counterclaim Entitlement

205. Consequential on the findings set out above, the total sums due to SRS by way of damages are as follows:

Head of Counterclaim	Sum Allowed
Prolongation – 22 Feb to 19 April	£105,088.50
Disruption – February to 18 April	£10,256.00
SMD disruption	Nil
SRS acceleration:	£522,659.05
A. Acceleration costs	£29,305.82
B. Additional site resources	£29,594.21
C. Senior management	
Mace Tower Crane cost	£95,692.69
Contra charges	£31,882.22
Total	£824,478.49

2.3.4.4 Mitigation

Lastly, some observations on 'mitigation' from HHJ Akenhead.

> 142. That said, I have no doubt and I find that it was sensible and reasonable for SRS to institute its recovery programme by way of mitigation to try to recover the delays totalling 42 days which had been caused by CBUK's breach of contract. It was reasonable because SRS was faced with a very substantial liquidated damages liability to Mace (£500,000 a week) and a six week delay (attributable to CBUK) could potentially have cost SRS some £3 million. To that one would have to add an additional six weeks' worth of site related and overhead costs which SRS would have had to incur. Therefore a concerted effort to work additional hours to try to recover the delay was at every level reasonable. Ultimately, it was not seriously argued otherwise on behalf of CBUK. I have regard to the authorities relating to mitigation of damage, one of the more well-known being Lord MacMillan's dictum in Banco De Portugal v. Waterlow & Sons Ltd [1932]
>
> 161. I refer to my findings above in relation to acceleration. It therefore falls to the Court to decide on the evidence what were the actual and the reasonable costs incurred by SRS in accelerating the work in what turned out to be an unsuccessful attempt to recover the delays which had occurred. Essentially, what happened was that from late February 2010, Steelcraft and its sub-sub-contractors began to work double shifts up to 11 pm with additional time over the weekend, this being implemented to seek to recover and mitigate those delays which had already been caused by the late deliveries of steel by CBUK; further night-time working began in or about March although it appears that at least initially proper consent had not been obtained from Southwark Council. There were noise trials carried out in April 2010 going into May and approval from the Council was received later in May. The recovery programme was initially based only on the extended hours (7 am to 8 am and 6 pm to 11 pm during weekdays and 7 am to 8 am and 1 pm to 4 pm on Saturdays). It is clear however that full night-time and weekend shifts were being worked over and above this recovery programme. It must be understood that working on a site such as this at night is and was necessarily less productive primarily because there is a need to keep noise down and steel erection can be a very noisy business; there were restrictions on using hammers at night. This meant that SRS could only fix steel members into position often only pinning them together with one bolt as a temporary measure until the connections were finally and fully fixed by the wrapping up team the following day. Remedial works could not be done at night if they required hammering.

169. I have formed the view that the reasonable costs of seeking to mitigate the impact of the culpable delays caused by CBUK can and should be assessed as follows:

   (a) MCS and K-Len were primarily deployed by Steelcraft with SRS' consent and knowledge to seek to overcome the delays to the steelwork erection which had started to be generated by the end of February 2010. By the time that the recovery programme was put in place in May 2010, that delay was 56 days or 8 weeks of which 6 weeks was attributable to the breaches of CBUK.

   (b) As from May 2010, these resources were deployed and were initially effective; indeed in June 2010 some time had been recovered. However, by September 2010 other factors had probably combined to cause further delay and as from the end of August 2010 these other factors represented the dominant reason or cause for the need to continue to retain these additional accelerative resources. Therefore as from then it cannot be said that there is any or certainly any sufficient causative link between CBUK's breaches and the need to continue to deploy the accelerative resources.

   (c) Mr Davis' approach based by reference to the recovery programme is a reasonable and realistic one, albeit that his exclusion of nightshift and his rates are unrealistic. It seems to me that, although the recovery programme only extended day shifts and Saturday working, SRS was always intending to and needed to deploy what it actually deployed which was full nightshifts and full weekend working as well to seek to recover the time which had been lost.

   (d) Therefore the proper amount to allow to SRS and what it was liable to pay its sub-contractors is reflected in the following table: ...

To this should be added Steelcraft's overhead and profit at 9% and 6% respectively, given that it was asked to provide the accelerative resources and that it was not to blame in any way for the six weeks delay caused by CBUK which those resources were deployed to overcome. This also reflects Steelcraft's entitlement to be paid a reasonable sum for providing the accelerative resource. This produces a total of £522,659.05.

2.3.4.5 Commentary

In the judgment, Mr Justice Akenhead gives some important observations regarding the presentation and supporting information necessary for disputes concerning disruption and loss of productivity.

With regard to 'disruption' and 'loss of productivity', Mr Justice Akenhead made several important observations in his judgment.

1. For his quantification of the disruption caused to SRS's works, their Expert was unable to use to use either of the two most recognised methods, the 'measured mile' or 'earned value' approaches. In this respect, Mr Justice Akenhead stated in paragraph 155:

> "He [the Expert] did not have appropriately detailed timesheets or daily record sheets to enable standing or non-productive time to be identified. He could not tell from the documentation whether the steelwork deliveries to site resulted in downtime and lost erection hours. He could not find a resource schedule or resource planning documentation within SRS or indeed Steelcraft to see what was allowed for. Finally, he says unequivocally in his first report at Paragraph 10.12.23 that he had not located detailed records for Steelcraft labour that would facilitate any, or any meaningful disruption analysis. He accepted in cross-examination that it was at least logically possible that the reason for the increased number of shifts was inadequate planning, inefficient use of cranes or even that Steelcraft was not very good at its job. There is limited reliable contemporaneous evidence that the so-called planned resource of 630 shifts was the planned resource; it was supported by Mr Willis but without reference to any contemporaneous document and it was contradicted by the original pleading which identified a higher planned resource."

2. On the matter of 'delay caused by crane downtime attributable to CBuk's late delivery of steel', HHJ Akenhead was not persuaded that this had been proven. He particularly noted that the Expert in his report stated, "that he had not been able to use any of the normal approaches to evaluating disruption. For instance he had not been able to use the 'measured mile' or 'earned valu'approaches".

However, Mr Justice Akenhead states:

> "Deliveries by CBUK were late and in some respects out of sequence. This led to delays and piecemeal deliveries to site. There must have been disruption in the sense of reduced productivity… What the Court can and should do in circumstances where it is satisfied on a balance of probabilities that some (more than *de minimis*) disruption must have occurred as a result of CBUK's breaches is to make a reasoned assessment albeit based on the minimum probably so attributable… Accordingly, I find that 40 man shifts' worth of disruption cost was incurred the rate established by Mr xxx (which I accept), namely £217.84, and that £8,713.60 consequentially represents the disruption labour cost attributable to CBUK's breaches of contract."

On the matter of 'acceleration', Mr Justice Akenhead expressed some displeasure over the Expert's lack of agreement, stating in paragraph 123:

> "Having agreed this, Mr Holloway then abandoned this sensible agreement, which was unfortunate and indeed unhelpful. Mr Barry, however, broadly stuck to the agreement and analysed the delay at least up until the commencement of Phase 103 at Levels 10/11. Neither expert either did or was readily able to analyse the delays to the remainder of the steelwork erection up to Level 40. This is because the history and the facts were all very confused."

Mr Justice Akenhead made the following observation in paragraph 143:

> "It follows that the reasonable costs incurred by SRS in implementing acceleration measures to overcome the 42 days delay caused to the steel erection programme by CBUK's failures to deliver steel in accordance with the December Programme should be recoverable. The fact that the acceleration was not ultimately successful due to factors which SRS can not establish are attributable to those failures can not and should not prevent or limit recovery of such costs."

On the matter of 'mitigation', Mr Justice Akenhead made an important observation in paragraph 169 regarding recovery of reasonable costs when attempting to mitigate culpable delays:

> "I have formed the view that the reasonable costs of seeking to mitigate the impact of the culpable delays caused by CBUK can and should be assessed as follows: 'xxx'."

Chapter 3
Planning, Programmes and Record Keeping

This chapter consists of five parts, namely:
  (a) Background and history of planning
  (b) Planning and programming
  (c) Programme submission, review and acceptance
  (d) Programme updates and revisions
  (e) Progress records and other record keeping.

3.1 Background and history of planning

From earliest recorded times groups of people have been organised to work together towards planned goals, and planners coordinated and controlled their efforts to achieve desired outcomes.

3.1.1 The early days

Considerable planning skills were required by, for example, the ancient Egyptians to build their pyramids, the ancient Chinese to build the Great Wall of China, and the Romans when building their roads, aqueducts and Hadrian's Wall.

These time-enduring construction projects required large amounts of human effort with planning, organisation and coordination; and all with no computers, faxes and the combustion engine!

After the fall of the Roman Empire, the great Dark Ages descended, and it was not until the mechanical clock and Guttenberg's moveable typefaces were invented that any further major development in 'planning' was forthcoming. The clock, invented by Heinrich von Wych in Paris in 1370 permitted accurate work measurement. The printing press allowed the ability to communicate by the printed word; and it was whilst at an early version of the Octoberfest that Guttenberg visualised the technique of combining the small dies used for

---

Practical Guide to Disruption and Productivity Loss on Construction and Engineering Projects, First Edition. Roger Gibson.
© 2015 John Wiley & Sons, Ltd. Published 2015 by John Wiley & Sons, Ltd.

coin-punching with the mechanics of a wine press. This produced a printed page, made up of moveable individual letters instead of from a single engraved block.

Developments in planning and production management then began. In 1436, a Spanish visitor to the Arsenal of Venice reported:

> "And as one enters the gate there is a great street on either hand with the sea in the middle. On one side are windows opening out of the house of the arsenal, and the same on the other side, and out came a galley towed by a boat, and in the windows they handed out to them, from one the cordage, from another the ballistics and mortars, and so from all sides everything which was required. When the galley had reached the end of the street all the men required were on board, together with the complement of oars, and she was equipped from end to end."

This was an example of the planning of a production line; half a millennium before Henry Ford!

It was not until the First World War that simple barcharts were used by the British army for planning military exercises. Then some 15 years later, construction began in 1930 of the Empire State Building, which on completion was the tallest building in the world and remained so for over 40 years. However, during its construction it was a marvel of programming excellence, the works being elaborately planned and programmed by Andrew Eken, the chief engineer of the contractor. The site in downtown Manhattan was very congested and there were virtually no lay-down areas. Material deliveries were carefully planned to coincide precisely with the installation works. Another impressive fact of this project is that the building's 56,000 tons of structural steel were erected in six months at the remarkable rate of 4.5 floors per week, all this without the aid of a CPM program or a computer!

3.1.2 Modern times

Mention 'planning' to the average person nowadays, and he or she will think of a barchart, which is the most common form of visually showing a project. The barchart is strictly speaking a Gantt Chart, named after the inventor, Henry Laurence Gantt, an American engineer. He is the first known person to publish a plan in a barchart format – and probably the first to be told off for a project not going to plan!

The Gantt Chart for which Henry will be remembered is a visual display chart used to present a schedule or programme of activities. It is based on time rather than quantity, volume or weight. In other words, a Gantt Chart is a horizontal bar chart that graphically displays time relationships. In effect, it is a 'scale' model of time because the bars are different lengths depending upon the amount of time they represent. Gantt charts have been around since the early 1900s and provide a method of determining the sequence and time to be taken to achieve a given objective.

3.1.3 The Critical Path Method (CPM)

The next major development in 'planning' was the advent of PERT and the emergence of Critical Path Method (CPM) programmes.

PERT is an acronym of 'Project Evaluation & Review Technique', and is a variation on CPM programmes that takes a more sceptical view of activity durations. To follow this technique, for each activity you estimate, (i) the shortest possible time the activity will take, (ii) the most likely length of time for the activity and, (iii) the longest time that might be taken if the activity takes longer than expected. Using the following formula, the duration for each activity is calculated,

$$\frac{\text{Shortest time (i)} + 4 \times \text{most likely time (ii)} + \text{longest time (iii)}}{6}$$

Using the PERT technique helps to bias activity durations away from the unrealistically short time scales sometimes assumed.

CPM programming has been around since the 1950s. The first known use appears to be in North America by E.I. DuPont Nemours Co, who developed a CPM programme in 1956 for construction of its $10 million chemical plant in Kentucky. The CPM was run on a large mainframe computer, called a UNIVAC, some 25 feet high and 50 feet in length containing 5000 tubes and 18,000 crystal diodes.

The type of CPM used in those days was the Activity Diagram Method (ADM) form of network.

However, the first high-profile use of a CPM was the Polaris project. This was in 1957, and the team on this huge project had to understand the development process for the most complex machine ever devised by man. They wanted a technique that would get the missile developed and into action. It was management consultants Booz, Allen and Hamilton that used the CPM to draw their maps of time. Polaris went on to hit the target of 'time'. Everyone celebrated the new behemoth, and the modern version of 'planning' was truly born.

3.1.4 The use of computers and planning software

CPM, preceded by its reputation, spread to other industries and to other environments. In the early 1970s, CPMs were run on main frame computer systems – however, few people owned these massive main frame computers on which the systems ran. The rest of us rented time at a now extinct breed of companies known as 'computer bureaux'. The procedure was that, after hiring time at a computer bureau that was running a critical path software program such as Projacs, we worked alongside a data entry person (and it was always a she in those days). From our data entry sheets, she would type an incomprehensible series of characters and numbers into on-screen forms which created a series of punch cards. These were fed into the main frame computer and processed. The result was a print-out of our project plan, often containing thousands of tasks or activities, and requiring long corridor walls for the huge print-outs of green and white stripy paper containing the network diagram. Computer people in the 1970s all had white coats, little white hats and a supercilious smile. Nowadays, only the white coats and hats have gone

In the 1980s, a great advance in computing took place in a garage in California; Steve Jobs was astounding the techie world with his Apple II. The Apple II, a small computer that sat on a desk-top, changed many peoples' lives. A huge push was given to the personal computer industry when IBM developed the PC. PCs sprouted everywhere and computer packages and planning software, such as MicroPert, were specially written for the new PCs. Critical path diagrams increasingly took a back-seat as much simpler barcharts were quickly and easily drawn by the new software.

The advent and development of personal computers and planning software packages specially written for them allowed the planning of a large construction project to be done on site. The other effect was that these affordable small computers and planning systems spread into other industries and onto smaller projects. Critical path diagrams and barcharts appeared on walls in offices throughout the world.

3.1.5 Precedence Diagram Method Networks (PDMs)

At the turn of the millennium, the Precedence Diagram Method (PDM) form of network supplanted the Arrow Diagram Method (ADM) form of network as the preferred planning method for CPMs.

The Precedence Diagram Method was developed in the early 1960s by an American company, H.B. Zachary, in conjunction with IBM. It was common in the 1970s and 1980s for planning software programs to accept and perform calculations for either ADM or PDM networks. However, from the 1990s new software was written only for PDM networks. For example, when Primavera software writers created a Windows version, they opted to use PDM as the platform for the program.

3.2 *Planning and programming*

Time is money; so the old adage says. However, poor planning/programming still ranks in most surveys as being within the top three problem areas that lead to project failure.

Therefore, the 'planning' of a project is a necessity for success; and one would expect that in this day and age with computers and planning software available to assist project managers and planners, delays would have been significantly reduced. However, results show the opposite.

Time is money; therefore planning shouldn't be ignored. Sadly, the statement that 'if you fail to plan, then you plan to fail' is often true, and sadly too many planners nowadays rely solely on a bit of computer software for 'planning'. The manner in which computers and planning software deals with activity logic and relationships through an interactive screen is an improvement from the days of creating a network program through punch cards. However, it also encourages planners to generate programs with illogical activity relationship links; and often this ill-conceived planning is a hindrance in time-related disputes.

Planning and programming are two separate functions, but are often linked together under the general term of 'planning'. However, before you prepare a programme, you must have a plan.

3.2.1 Planning

To plan a project means to identify the tasks or work activities to be performed and logically inter-relate them. The questions of time for performance and resources required are answered as part of the programming function.

The first stage is a broad-brush approach, and it is best to start with a blank piece of paper – not a computer.

For example, take a six storey concrete-framed commercial building. First, assess how long it will take for the main elements, i.e. (i) substructure works before starting superstructure work, (ii) superstructure work, (iii) envelope and cladding, (iv) services works, and (v) internal finishes.

Let's say the three main tasks in the substructure work are bulk excavation, piling, and concrete works (being pile caps, ground beams and ground floor slab). Taking a broad-brush approach, these are assessed as 2 weeks, 4 weeks and 6 weeks respectively. However, the planner, using his experience, allows for overlapping between these main stages and his conclusion is that work to the concrete frame superstructure can start 10 weeks after starting the bulk excavation.

The next key element, superstructure, is approached in a similar fashion. However for this element the key is the cycle time for a typical floor of $400\,m^2$. Now the planner has to go into more detail to assess how long for fixing the formwork and reinforcement for this area of concrete slab. He also has to take into consideration crane hook-time; that means how many lifts for the single tower crane positioned in the central core area. For the example we are using, it is probably the crane hook time that is the governing factor when assessing the cycle time for a typical floor.

On a construction project, 'planning' covers all aspects, from overall planning– building 'A' must be completed before building 'B' can start – down to detailed planning – the activity 'excavate for foundations' has to be completed before its successor, 'pour concrete in foundations', can start.

By planning the works in detail, and linking activities in a logical manner, a contractor creates a network of activities and their dependencies or inter-relationships as shown above. If this is done in a proper manner encompassing all works and all restraints on the project, then this is the basis for a critical path network.

The next stage is to calculate the time each activity will take. This phase is the start of preparing the programme for the project. For example, for 'excavate for foundations', the contractor will know he has 1000 cubic metres of soil to dig out, and at a productivity rate of 100 cubic metres per day this activity will take 10 days. This is known as the activity's 'duration'.

After completing this exercise for all activities, he then has a 'time frame' for the project. For example, 'excavate for foundations' will start on day 1 and because it has duration of 10 days, it will finish on day 10. Its successor, 'pour concrete in foundations', will start on day 11 and as it has a duration of, let's say 15 days, will finish on day 26. The contractor now has a programme.

3.2.2 Programming

In its simplest terms, programming (or 'scheduling' as it's sometimes called) is a method whereby the work activities necessary to be performed in order to achieve project completion are arranged in a logical order.

A properly developed programme will not only show the sequence in which the activities are intended to be carried out, but will also enable the participants of the project to monitor progress. In addition, the programme will be able to project future work while providing historical data that could be useful in analysing the past. This most common type of programme is a bar chart; either hand drawn or, more likely nowadays, computer-generated using commercially-available project planning software.

3.2.3 CPM programmes

A critical path method (CPM) programme refers to the development of a logically-linked network that enables the identification of a critical path. The critical path is the longest activity path from the start of the project to its completion.

Activities on the critical path have no float; conversely, an activity not on the critical path will have float. Float is the amount of time an activity can be delayed without it becoming a critical path activity. Any activity on the critical path that experiences a delay will consequently delay the project completion date.

The calculations necessary to determine activity start and finish dates, together with float and the identification of the critical path, are very simple. These CPM calculations can be performed manually, but with computers and project planning software the thousands of calculations representative of a typical construction project can be effectively compiled and organised into an intelligible format in a matter of seconds.

A network programme, or CPM, provides the ability to analyse the effect of every activity on the project completion date, and far outstrips other methods for progress monitoring, reprogramming, or evaluating new factors. At any time, you can determine if an activity is or is not on the critical path, and whether there is any float associated with that activity. If there is float, you will know precisely how much that activity can be delayed without it impacting the project completion date. This knowledge enables the project team to track and control progress, and to mitigate delay to the project completion date should critical activities be delayed.

However, a word of warning. The level of detail that will exist in a CPM programme is largely a matter of judgement on the part of the planner. Too much detail could conceal significant factors, whilst too little detail may result in a programme that is not very meaningful.

3.2.4 What is the use or benefit of a CPM programme?

By preparing a CPM, a contractor reassures himself that he can complete all the works and achieve completion of the project by the contract completion date. He knows when he has to have available key resources or equipment. Using the earlier simple example, he knows that he is going to 'pour concrete in foundations' starting on day 11, therefore he will have to have his concrete-producing equipment up and running by this date.

The benefit of a CPM programme for the employer or contract administrator is that they are also reassured that the contractor can complete the project on time, and that he has planned the works in a reasonable and logical manner. Again using the earlier example, the employer knows at an early date that the contractor intends to start 'pour concrete in foundations' on day 11 and that he has to provide the drawings for this work before this date.

3.2.5 Pitfalls in the use of CPM

Although the above sections extol the virtues of CPM, one should be aware of the associated inherent dangers. Detailed below are four of the most common pitfalls.

3.2.5.1 Quality of the CPM programme

Readily available user-friendly project planning software makes it possible for almost any computer-literate person to create a CPM programme that appears to be reasonable. It is very easy to input various activities that comprise a project into the software and string them together in such a way that, when looking at a barchart print-out, the work seems to flow in a way that seems entirely sensible. Unfortunately, there is no way to tell simply by looking at the printed barchart whether this is a true CPM programme, or simply an 'attractive barchart'. Very often, no network logic or activity relationship links are issued with the programme, and therefore the barchart print-out on its own is essentially useless.

Flawed programme logic can be hidden from all project participants unless someone works directly with the project planning software on the computer. Without this direct examination of the electronic file, the programme may, either intentionally or unintentionally:

- contain flawed logic
- include activity constraints that interrupt the calculation of the critical path and/or floatshow only those activities that the contractor wishes the employer or contract administrator to see
- misrepresent project status at a progress update.

Therefore, the only way to avoid these circumstances is to require electronic copies of the baseline, or as planned, programme and all subsequent updates to be submitted with the barchart printout and paper reports.

3.2.5.2 What is critical today may not be critical tomorrow

The critical path identified in the original baseline, or as planned programme, will only remain the critical path if everything goes according to plan. As everyone knows, this is almost never the case, as the calculated project completion date is directly dependent upon the completion of every activity on the critical path taking no longer than originally estimated.

Furthermore, if an activity not on the original critical path is delayed by more than its available float, then it will become critical and, in effect, the project's critical path has changed.

3.2.5.3 Unrealistic activity durations

For many programme activities that are delayed the as-planned duration may have been entirely appropriate, but this is not always the case. All too often the duration of activities in a CPM programme are wild guesses that are unrealistically short or, in some cases, excessively long. All activities on the critical path and those that are near critical should have supporting data to show how their durations have been calculated.

3.2.5.4 Managing the programme and regular updates

Unfortunately, many contractors view a programme as nothing more than a requirement of the contract, and do not take it seriously enough to properly develop a CPM programme and maintain it as a management tool. Without proper attention the CPM will become nothing more than a list of activities and a convenient way to record actual start and finish dates.

One of the principles to be followed in maintaining a CPM programme is the regular monitoring of the work by a periodic review of the programme. Programme updates should be performed on a regular basis for the purpose of gathering progress information and revising programme logic as appropriate. The project planning software takes this contemporaneous information and recalculates the critical path so that management knows which activities are now driving the project completion date. The update also records project history, as well as projecting start and finish dates for future activities. If the CPM programme is not updated on a regular basis, it will quickly become inaccurate and consequently useless.

This update information must be collected, inputted and analysed relatively quickly so that the update reports can be distributed to the project's participants while there is still time to react. A CPM programme is dynamic in nature and the critical path is continually evolving over time. Failure to disseminate the update information in a timely fashion may render the information useless from the standpoint of being able to proactively manage the project.

3.2.6 Types of 'programmes'

The construction industry uses a number of different types of programmes to manage projects. There are two types of programmes in most common use: the 'barchart' and the 'network'.

The most frequently-used of these two programmes is the barchart, which is a list of those activities for the project. The planned start and planned finish of each activity are shown in a time grid and are connected in a bar. The bar therefore represents the duration of the activity. The assumption usually made is that the bar represents a continuous interrupted activity, but this may not be the case. It is usual to include a tabular listing of the activities on the left-hand side of the barchart, which may include calendar start and finish dates together with overall durations for each activity.

An example of a simple barchart is given in Figure 3.1, showing the construction of a garage.

The barchart is easily prepared and can be used to show estimated timing and duration of activities, and to record actual progress. It does not require special software or computers and can be drawn easily by hand. The types of activities

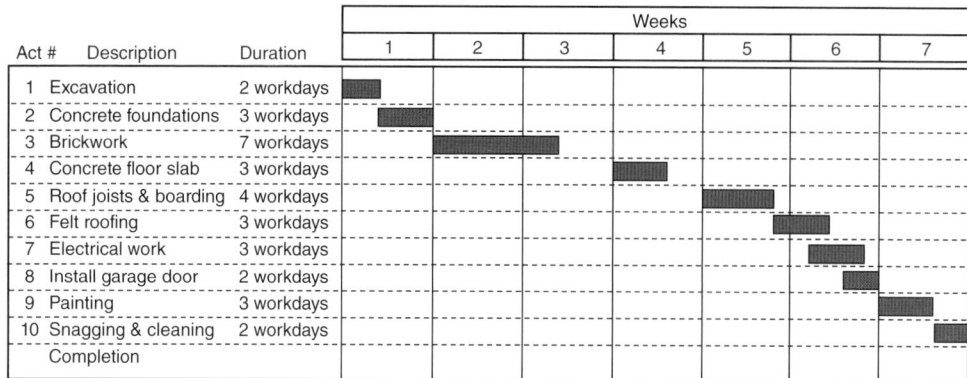

Figure 3.1 Example of a simple barchart.

are not limited in any way, since the barchart is simply a diagrammatic representation of the time characteristics of an activity.

The barchart does not model the inter-relationship between activities, and does not model the consequences on the project completion date if the actual timing or duration of an activity is not met. So, for example, if an activity is started later than shown on the programme, the barchart does not allow the effect on completion to be analysed, without additional information. The barchart simply shows that an activity started later than planned. Similarly, if an activity takes longer than its duration as shown on the programme, the barchart only shows that an activity took longer than estimated. The barchart therefore simply provides a model of the time characteristics of the activity, and does not model the relationship of the activity with the time characteristics of the project and, in particular, the project completion date.

The absence of logic links between activities means that the use of barcharts is limited to monitoring progress rather than forward planning of the project. It is often used in the initial stages of delay analysis to compare planned and actual progress so as to identify problem activities. Care is required, since the implicit assumption that the planned durations were an accurate and still valid estimate may not be correct.

The second type of programme, the network, is a model not only of the activities and their durations, but of their interdependence and association with completion of the project.

The most used network programme is the Critical Path Method (CPM) which models the construction logic links between the activities. The construction logic represents those factors which define the construction sequence of the project and include:

- the method of working; showing how the project is to be carried out and the sequence of activities;

- the construction constraints; which may be access dates for parts of the site or release dates for information or delivery dates for work by others.

There are two types of CPM programmes; Activity-on-Arrow and Activity-on-Node.

The Activity-on-Arrow programme produces an arrow network in which each node represents either the beginning or end of a discrete activity and the arrow linking the nodes is the activity. The nodes are numbered and the activity is identified by the numbers of the nodes at the start and the finish. The example in Figure 3.2 shows the activities for the construction of a garage (as used in the barchart programme example in Figure 3.1). Activity "Excavation" is identified as Activity 21–22. Node 21 represents the beginning of Activity 21–22 and node 22 the end of Activity 21–22. Node 22 is also the start of Activity 22–23. This demonstrates the logic inherent in the Activity-on-Arrow which is *finish-to-start*. The arrow activity is not drawn to a time-scale but the duration is annotated as shown in the example. The Activity-on-Arrow network is useful in representing the flow of work, but its use has declined in construction.

In the Arrow-on-Node programme, each node is an activity with a duration and the arrows represent the logic link between the activities. The programme uses *finish-to-start* relationships or links which are the same as used in Activity-on-Arrow programme. In the example in Figure 3.3 for the construction of a garage, Activity 36 is "Felt Roofing" and is linked to Activities 37 and 38. Activity 38 can start once Activity 36 is completed, whilst Activity 37 can start two days after Activity 36 has started and can only finish one day after the finish of Activity 36. The Precedence Network Method is now the most common form of Activity-on-Node programme and uses the possibility of defining the links between activities by relationships other than *finish-to-start*. This method permits not only *start–finish* links, but *start–start* and *finish–finish*, as well as allowing a time dimension to be added to the link in the form of a *lag* or *lead*. The choice of logic link depends on which link accurately models the particular restraint.

The facility to define the relationship of activities, both in terms of the type of logic as well as with a time dimension, makes the Precedence Network Method a most powerful and flexible method of programming. The assumptions made must be carefully examined when carrying out any delay analysis or management through programming analysis. If, for instance, the initial design of the equipment in the above example is delayed, then the lag in the above *start–start* link will need to be adjusted to take account of the delay. In any analysis, the time dimension of links which are not based on real time factors needs to be examined carefully to establish that it still accurately models the relationship between activities.

Nowadays, it is common for the network to be presented as a time scaled logic linked programme, as shown in Figure 3.4.

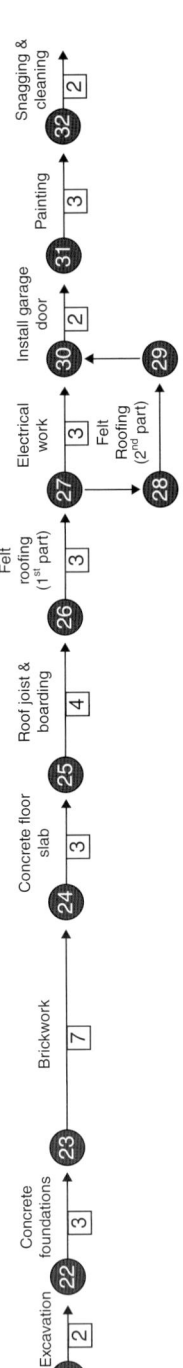

Figure 3.2 Activity-on-Arrow network.

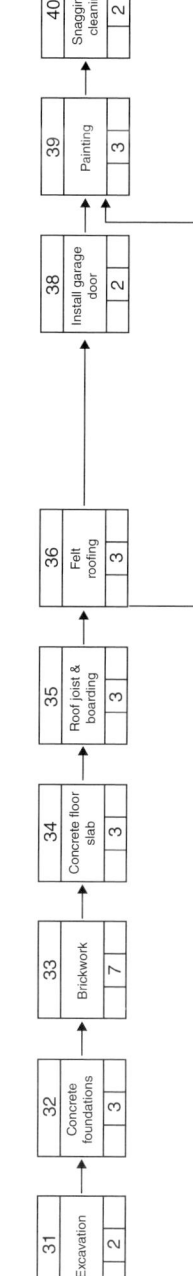

Figure 3.3 Activity-on-Node network.

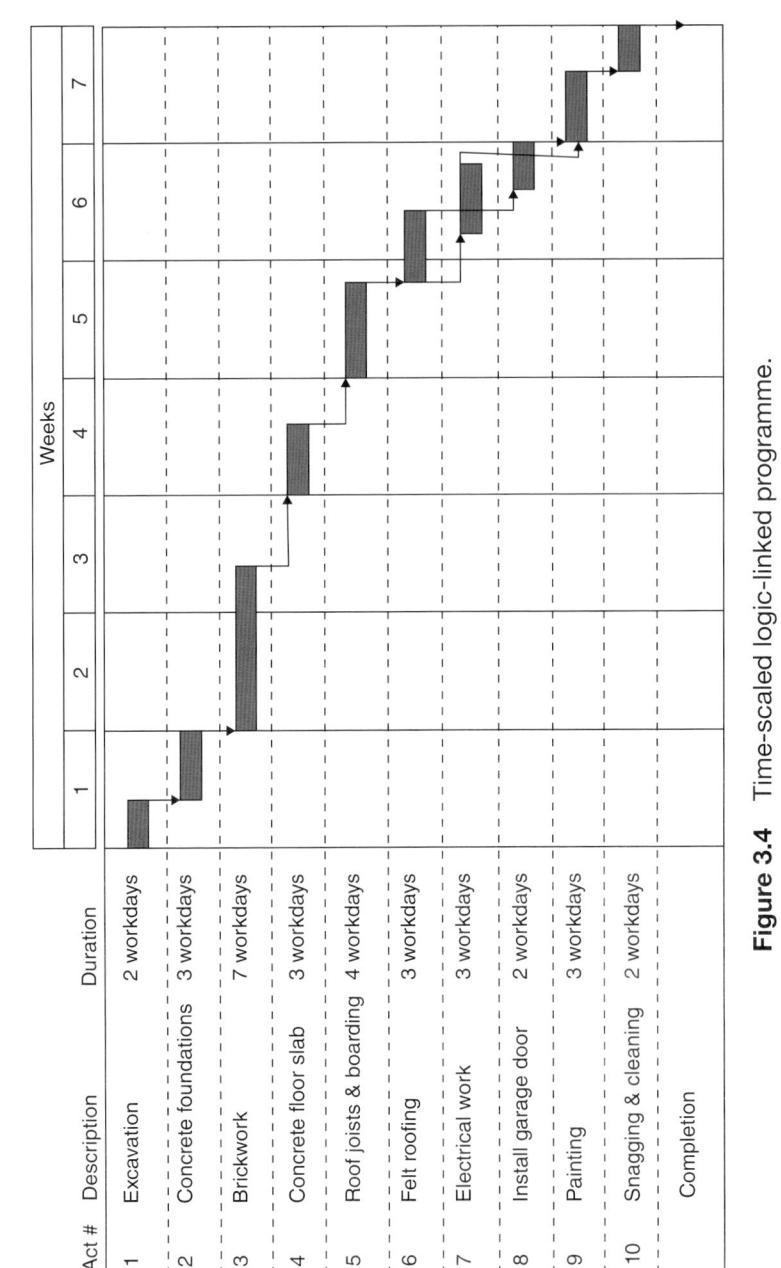

Figure 3.4 Time-scaled logic-linked programme.

In order to make management decisions and to establish the priority of actions, it is necessary to interrogate the network.

One important attribute which is relevant to the obligation to complete by a specified date is the critical path. Those activities which can be least delayed without affecting the date for completion are said to be on the critical path. The line, or path, through those activities is the critical path to completion and is usually generated by modern software

This is shown in Figure 3.5 for the garage construction project, with the activities on the critical path shown as bold arrows.

The activities which are not on the critical path will have 'float'. This is usually shown as the difference between the earliest and latest start dates. There are various types of float, all of which are an expression of the relationship of an activity to other activities and milestones. The term 'float' used here is the period by which an activity on a programme may be delayed before the programme shows an effect on the date for completion. The activities with the least float are on the critical path to completion.

The emphasis on the programme is important because float is a function of the model represented by the programme, but may not accurately represent the consequences of starting an activity later than it could have been started. The construction logic on many programmes is kept simple in order to produce a workable programme so that management decisions can be taken.

Two other types of programmes are 'Line of Balance' and 'Time Chainage'.

3.2.6.1 Line of balance

The main concept of the Line of Balance technique is the work continuity of labour gangs, which work with rhythmic production and no wastes are willingly planned into their programme. This planning method fits much more closely to modern construction philosophy.

The Line of Balance technique is very suitable for repetitive projects like residential buildings; however, it may be adapted for non-repetitive projects as well. Unlike a barchart, which shows the duration of a particular activity, the Line of Balance chart shows the rate at which the work activity or group of activities have to be undertaken to stay on programme, and it shows the relationship of one activity or group of activities to the subsequent group. More importantly, it shows that if one group is running behind programme, it will impact on the following group.

The main advantage of this technique is its graphical presentation and easy understanding of the programme,.

Figure 3.6 is a sample Line of Balance chart for a residential development of 20 houses. In the chart, the 'x' axis is the 'time' (in this example expressed as working weeks), whilst the 'y' axis is the number of houses.

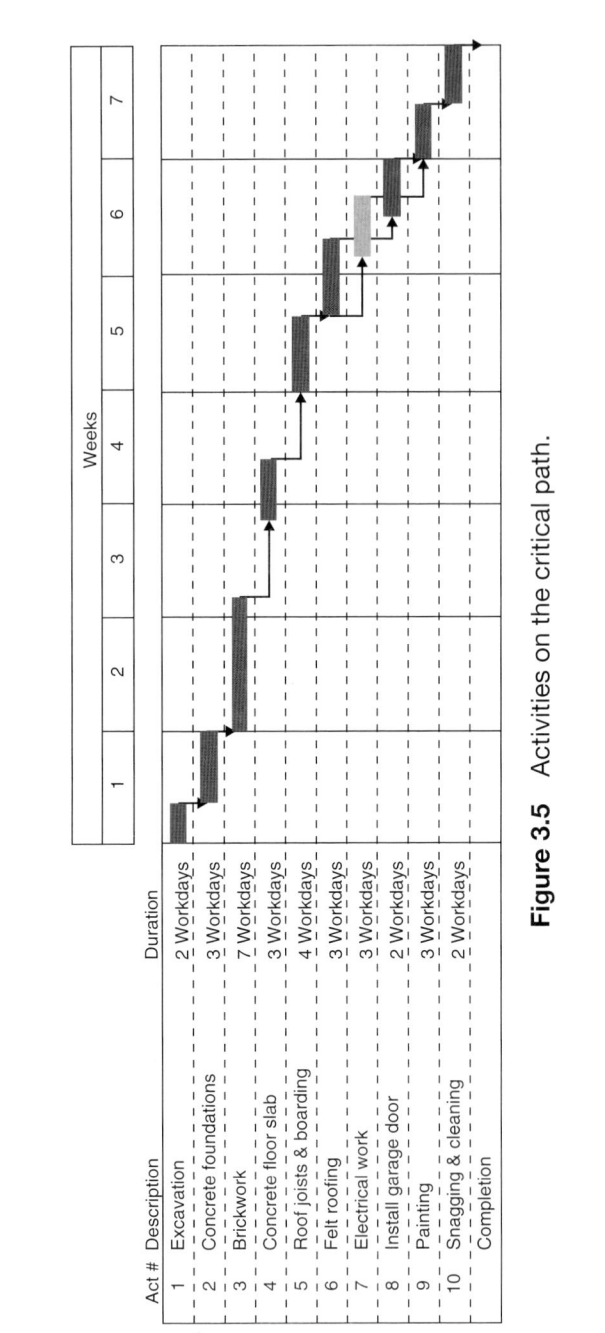

Figure 3.5 Activities on the critical path.

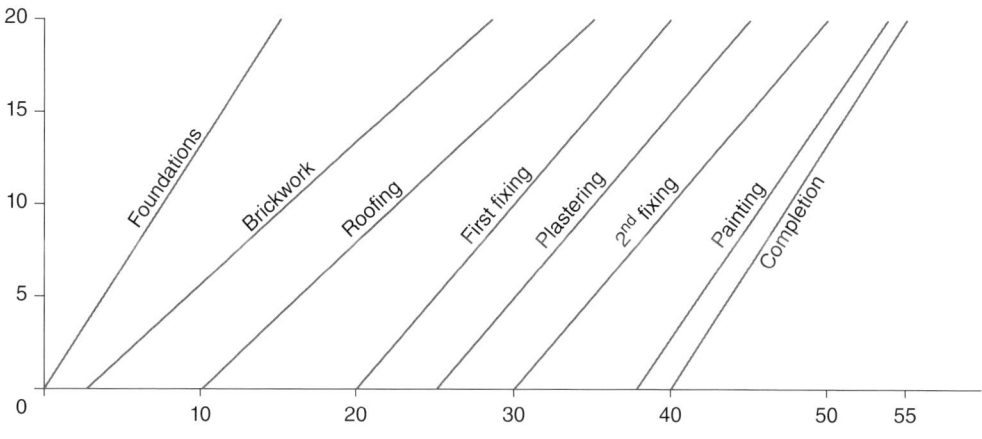

Figure 3.6 A Line of Balance chart.

3.2.6.2 Time Chainage

For certain types of projects Time Chainage provides a clearer, more easily understood picture of the plan than the traditional barchart because it has a more graphical structure. The types of project that lend themselves to Time Chainage programming technique are:

- Roads
- Railways
- Pipelines
- Tunnels
- Transmission lines.

Time is displayed on the 'x' axis and distance is displayed on the 'y' axis. The chart shows the planned start and finish of a work activity against the actual location, or chainage, it is operating within the site as a diagonal line. The chart will also show fixed structures, such as bridges and culverts, as block sections for a fixed period of time.

3.2.7 Levels of programmes

Planners often describe the various types of programmes/schedules they produce as being of different levels. Each individual person or organisation would set up and use their own system for describing these various levels of programmes and this has led to confusion caused by inconsistency.

It is time that the system for describing this hierarchy is standardised in order that some consistency is achieved, so that people can understand what is being

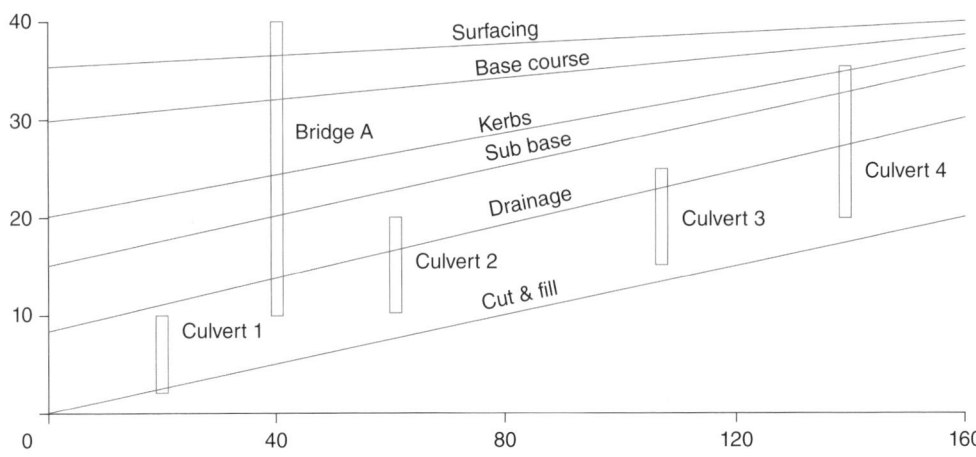

Figure 3.7 A Time Chainage chart.

referred to by, say a level 2 programme. Therefore, recently the *Planning Engineers Organisation* produced a paper to set out the standards of description that all planners and schedulers can use. The purpose of this paper is not to determine or set out what programmes / schedules should be produced by whom at what stage. Its use should be limited to the standardisation of the terminology given to each level of programme / schedule. It is hoped that the recommendations of the *Planning Engineers Organisation* will be adopted by planners and schedulers as being the reference standard against which in future, programmes / schedules can be described.

The *Planning Engineers Organisation* have kindly given permission for this paper to be reproduced and included in this book, and this is included as Appendix 2 towards the end of the book.

3.3 *Programme submission, review and acceptance*

There is a clear need for a 'baseline' programme to be developed after the award of contract, reflecting the intentions of the contractor.

Contract administrators need front-line skills to review a contractor's baseline programme. Accordingly, contract administrators increasingly have to decide if, and to what extent, they are going to trust, approve or accept a contractor's programme submissions. In today's planning software paradise, CAs should be able to detect common techniques or mistakes when reviewing programmes could increase the likelihood of extension of time awards. These techniques mean that a programme will not function as a proper predictive tool for measuring progress or quantifying the impact of delays and changes.

3.3.1 Contract requirements: JCT 2005

The Joint Contracts Tribunal Standard Building Contract, With Quantities, 2005, includes in Section 2 Clause 2.9, the following:

Construction information and Contractor's master programme

2.9.1 As soon as possible after the execution of this Contract, if not previously provided:
1. the Architect/Contract Administrator, without charge to the Contractor, shall provide him with 2 copies of any descriptive schedules or similar documents necessary for use for carrying out the Works (excluding any CDP Works); and
2. the Contractor shall without charge provide the Architect/Contract Administrator with 2 copies of his master programme for the execution of the Works and, within 14 days of any decision by the Architect/Contract Administrator under clause 28.1 or of any agreement of any Pre-agreed Adjustment, with 2 copies of an amendment or revision of that programme to take account of that decision or agreement.

But nothing in the descriptive schedules or similar documents (or in that master programme or in any amendment or revision of it) shall impose any obligation beyond those imposed by the Contract Documents.

3.3.1.1 Commentary on the JCT 05 requirements

JCT 05 has a very basic requirement for submittal of the contractor's programme, the only requirement being a '*master programme for the execution of the Works*'. Unlike the NEC3 Contract, there are no requirements on the content of the programme and supporting information.

3.3.2 Contract requirements: NEC3

The Engineering and Construction Contract, 'NEC3', includes in core clause 3, '*Time*', the following clauses:

'The programme 31

31.1 If a programme is not identified in the Contract Data, the Contractor submits a first programme to the Project Manager for acceptance within the period stated in the Contract Data.
31.2 The Contractor shows on each programme he submits for acceptance
- the starting date, access dates, Key Dates and Completion Date,
- planned Completion,

- the order and timing of the operations which the Contractor plans to do in order to Provide the Works,
- the order and timing of the work of the Employer and Others as last agreed with them by the Contractor or, if not so agreed, as stated in the Works Information,
- the dates when the Contractor plans to meet each Condition stated for the Key Dates and to complete other work needed to allow the Employer and Others to do their work,
- provisions for
  ○ float,
  ○ time risk allowances,
  ○ health and safety requirements and
  ○ the procedures set out in this contract,
- the dates when, in order to Provide the Works in accordance with the programme, the Contractor will need
  ○ access to a part of the Site if later than the access date,
  ○ acceptances,
  ○ Plant and Materials and other things to be provided by the Employer and
  ○ information from Others,
- for each operation, a statement of how the Contractor plans to do the work identifying the principal Equipment and other resources which he plans to use and
- other information which the Works Information requires the Contractor to show on a programme submitted for acceptance.

The next sub-clause, 31.3, concerns acceptance of the contractor's programme by the project manager, whilst clause 32 is titled *'Revising the programme'*. Both clauses 31.3 and 32 are referred to in Chapter 9 of this book.

3.3.2.1 Commentary on the NEC3 requirements

The NEC3 contract recognises that the programme is an important tool for use by both the contractor and project manager. The programme is valuable not only as a scheduling tool but also as a project management and change control tool.

NEC3 has distinctive features on the content of the contractor's programme. Indeed, the programme is the contractor's programme and he owns the terminal float. The programme is not only used to portray how the contractor intends to carry out the works, but can also be used for forensic analysis to determine the effect of compensation events for both time and money.

One of the key features of the programme under NEC3 is that, upon its acceptance, the contractor's programme becomes the *'Accepted Programme'*. Any subsequent programmes submitted by the contractor and accepted by the project manager in turn become the *'Accepted Programme'*, superseding the previous programme.

With regard to the required content of the contractor's programme, here are some matters to be aware of:

(i) *'Planned Completion'* is the date when the contractor plans to complete the works. The requirement is to show on the submitted programme both the *'planned Completion'* and the *'Completion Date'*. At the start of the contract the contractor's *'planned Completion'* may be a date earlier than the contractual *'Completion Date'*.

(ii) *'The order and timing of the operations which the Contractor plans to do in order to Provide the Works'*. This should be clear from the programme, i.e. network logic and listing of activities with start and finish dates. However, incompatibility in this document and with other contractor documents is sufficient reason for the Project Manager not accepting a programme. The requested information will also facilitate the assessment of compensation events. This item can also include such off-site manufacturing of components such as bathroom pods and the like. It is advisable that the procurement chain of these items, e.g. design, approvals, manufacture, etc, be included.

(iii) *'A statement of how the Contractor intends to do the work'*. In effect this is a resource statement, i.e. for each activity a list of the resources that are intended to be used. Clearly this list will be based on the scope of the work at the time of submittal of the programme. The resource statement will also facilitate the assessment of compensation events.

(iv) *'The order and timing of the work of the Employer and Other'* The employer and project manager need to ensure that any constraints on how the contractor is to 'provide the works' are stated in the 'works information'. The contractor needs to show these constraints in his planning and programme submittal. To introduce constraints at a later date, after commencement of the works, would be a change to the 'works information' and probably a compensation event.

(v) *'Provisions for float'*. This is an important aspect of NEC3, in that it recognises float in a programme. There are three types of float that should be addressed here:
 1. Terminal float: the period of time between the *planned Completion* and the *Completion Date*.
 2. Total float: the amount of time a programme activity can be delayed without affecting the *planned Completion* and reducing the terminal float.
 3. Free float: the amount of time a programme activity can be delayed before affecting a successor programme activity and thereby possibly reducing the total float.

(vi) *'Provisions for time risk allowances'*. Another important aspect of NEC3. An example of this is the amount of down-time allowed for an earthworks activity being carried out during winter. Time risk allowances are owned by the contractor and will be included in the Accepted Programme.

(vii) *'Acceptances'*. An example here is where the contractor is designing part of the works. If so, he should show on his programme the date(s) on which he requires acceptance of his design.
(viii) *'Plant and Materials and other things to be provided by the Employer'*. The contractor should show on his programme the dates when he requires plant and materials supplied by the employer.

3.3.3 What to look for in a programme review

When the programme is submitted, the CA should ask the following questions:

1. Does it comply with contractual obligations, milestones, or restraints on working hours or methods?
2. Is the entire scope of the work represented?
3. Are any activity durations questionably too long or too short for the scope of work they represent?
4. Are there any obvious errors in the programme related to the sequence or timing of the works?
5. Are there any onerous requirements of the employer's professional team, e.g. early completion programmes, unrealistic time allowances for approvals or supply of information, which are employer's risks?

3.3.3.1 Review of a CPM programme submittal

A very dangerous misunderstanding exists with a CPM programme submittal; many contract administrators and other professionals are still of the mistaken opinion that a CPM submittal consists of several pages of activity listings and/or a barchart plot or two. A CPM submission for review should consist of a full copy of the computer files necessary to recreate the programme; everything else is just frills.

A CPM submission, both for the baseline for review and subsequent updates, should consist of three discrete items, which are:

1. The activity details, including description, original and remaining durations, and percent complete. In conjunction with this, you should see for each activity other computed information such as early and late start and finish times, and total float.
2. The logical relationships that connect the various activities together to form a network which makes the CPM work. Full details of any lags and leads, i.e. imposed time durations between activities, is a must in the submittal.
3. Lastly, and certainly not least, is 'constraints'. The true logic of a network can be overridden by the programme containing various time constraints on

an activity(s). These will artificially reduce total float and could create an invisible delay, or even have the activity just expand to take all available time. This will never show up on a barchart plot and is only found in a 'constraint' listing and/or a copy of the computer files.

Having been satisfied that the information in the contractor's submittal is sufficient for a proper review, here are five basic checks or tests that should be carried out using the computer files provided by the contractor.

Test 1: Does the 'longest path' filter identify a reasonable critical path for the project?
Make sure the longest path is reasonable, and then check the reasonableness of near critical paths.

Test 2: Are there any open-ended activities in the programme?
In general, there should be only two open-ended activities in the entire network: one beginning activity with no predecessors, and one completion activity with no successors. Every other activity should be logically tied into the network. Furthermore, every activity should have its finish constrained with at least one FS (finish-to-start) or FF (finish-to-finish) successor relationship to another activity. Likewise, every activity should have at least one SS (start-to-start) or FS (finish-to-start) predecessor relationship to another activity.

Test 3: Do any of the activities have too much float?
Activities with too much float may indicate missing logic links, or logic links that have been overridden in a subsequent progress update. Identify any such activities.

Test 4: Are there any unnecessarily long gaps in workflow when grouping activities by work area and sorting by early start dates?
In most cases, once work begins in a particular area or phase of the project then the programme should allow work to continue uninterrupted in that area or phase. Long calendar gaps in a work area or phase may indicate less than ideal workflow and suggest an adjustment of preferential logic links to create a better plan.

Test 5: Are there activities with unnecessary contractor assigned constraints?
As constraints override the network logic in calculating activity start/finish dates and total float they should be used sparingly, if at all. A better approach is to use activity durations and network logic to model the project, and thereby eliminate constraints.

3.3.4 Acceptance of the programme

If the contract administrator fails to comment it may be implied as acceptance that the contractor's programme is contract compliant/satisfactory. When 'accepting' a programme, the contract administrator could be merely acknowledging receipt of the contractor's intentions. In 'approving' the programme, the contract administrator is more often seen to have performed some level of due diligence on the programme, such as asking the questions above, and is therefore acknowledging that the submission complies with the terms of the contract. However, it is important that a realistic baseline is established for the management of the works and the assessment of potential and actual effects of changes, unforeseen events or other circumstances that could delay the works.

Programmes are key documents in extension of time and delay claims disputes; therefore their significance in potential dispute resolution forums cannot be underestimated. At the same time, the perspective must be maintained that the programme is a management tool to assist in managing the work. A balance should be struck between keeping the contractor on an accurate progress path and the emphasis on the programme as a claims document. If approval is granted, this should not in any way relieve the contractor from complying with the contract, or in any way increase the employer's liability.

3.4 Programme updates and revisions

Notwithstanding the acceptance and popularity of detailed programmes, progress updates and their analyses in the dispute resolution arena, they are not held in the same esteem by many of the personnel on the project actually executing the work.

Criticisms that one hears on a construction project regarding programmes are either founded on the detailed use of the tool or the very output. Here are some typical criticisms of programmes and progress updates by site based personnel:

- The programme activities and CPM network are too detailed or too condensed.
- There is no feedback/dialogue between planner and site.
- The programme is difficult to read or be understood by those who need to use it.
- Activity durations are haphazard and often changed in subsequent progress updates without rationale.
- Programme updates schedules are out of date by the time they are issued.

3.4.1 Contract requirements: JCT 2005

The Joint Contracts Tribunal Standard Building Contract, With Quantities, 2005, contains no specific requirements for programme revisions, other than the following reference in Section 2 Clause 2.9:

2.9.1.2 the Contractor shall without charge provide the Architect/Contract Administrator with 2 copies of his master programme for the execution of the Works and, within 14 days of any decision by the Architect/Contract Administrator under clause 2.28.1 or of any agreement of any Pre-agreed Adjustment, with 2 copies of an amendment or revision of that programme to take account of that decision or agreement.

3.4.1.1 *Commentary on the JCT 05 requirements*

JCT 05 has only a reference in clause 2.9, for the contractor to submit a revised programme after a clause 2.28.1 decision or agreement of a *Pre-agreed Adjustment*. Clause 2.28 is titled, '*Fixing Completion Date*', and sub-clause .1 concerns the granting of an extension of time for a delay caused by a Relevant Event which is likely to cause delay in completion of the Works or a Section.

3.4.2 Contract requirements: NEC3

The Engineering and Construction Contract 'NEC3', includes in core clause 3, '*Time*', the clauses concerning the content, submission and acceptance of the contractor's programme. Sub-clauses 31.1 and .2 refer to the submission and content of the programme, whilst sub-clause 31.3 concerns approval of the programme as follows:

The programme 31

31.3 Within two weeks of the Contractor submitting a programme to him for acceptance, the Project Manager either accepts the programme or notifies the Contractor of his reasons for not accepting it. A reason for not accepting a programme is that
- the Contractor's plans which it shows are not practicable,
- it does not show the information which this contract requires,
- it does not represent the Contractor's plans realistically or
- it does not comply with the Works Information.

Clause 32 concerns revising the programme, and reads as follows,

Revising the programme 32

32.1 The Contractor shows on each revised programme
- the actual progress achieved on each operation and its effect upon the timing of the remaining work,
- the effects of implemented compensation events,
- how the Contractor plans to deal with any delays and to correct notified Defects and

- any other changes which the Contractor proposes to make to the Accepted Programme.

32.2 The Contractor submits a revised programme to the Project Manager for acceptance
- within the period for reply after the Project Manager has instructed him to,
- when the Contractor chooses to, and in any case,
- at no longer interval than the interval stated in the Contract Data from the starting date until Completion of the whole of the works.

3.4.2.1 Commentary on the NEC3 requirements

The contractor submits his programme and upon acceptance of the programme by the project manager it becomes the Accepted Programme. Subsequent programme submissions by the contractor become the Accepted Programme when accepted by the project manager.

The first programme is submitted with the tender or within a stipulated time, e.g. four weeks, after contract award. If the contractor does not submit his first programme within the time required, the project manager retains 25% of the value of the work done to date by the contractor until the first programme is submitted. This emphasises the importance placed on the programme by NEC3.

The contract gives only four reasons for which the project manager can refuse acceptance of the contractor's programme. These are:

1. *'The Contractor's plans which it shows are not practicable'*. The reason refers to the contractor's plans only. As an example, the contractor's original plan and programme for plasterwork shows an output of 100 sq.m. per gang per day, and his subsequent programme and plan shows 180 sq.m. per gang per day; whilst his actual production is currently showing 70 sq.m. per gang per day. His plans are therefore not realistic or *practicable*.
2. *'It does not show the information which this contract requires'*. This reason refers to the Contract, and it should be remembered that this also includes the *Works Information* and whatever else has been incorporated into the Contract. An example of this condition not being fulfilled is where the contractor's programme does not show *Key Dates* or *access dates*.
3. *'It does not represent the Contractor's plans realistically'*. For example, the contractor's programme is based on bored piles being used, whereas the project manager knows that driven piles have been procured by the contractor and are to be used.
4. *'it does not comply with the Works Information'*. An example may be that the contractor's programme has not taken into account a design constraint as shown on the *Works Information*.

If the project manager does not accept the programme then the contractor is obliged to resubmit the programme within the period for reply.

The contract gives the following reasons for which the contractor is to submit a revised programme to the project manager for acceptance. These are:

1. *'Within the period for reply after the Project Manager has instructed him to'.* If the project manager instructs the contractor to provide a revised programme for his acceptance then the contractor must do so.
2. *'When the Contractor chooses to'.* The contractor may choose to submit a revised programme to the project manager for acceptance. An example of this may be after deciding to change the sequence or method of working as stated on his then current *Accepted Programme.*
3. *'At no longer interval than the interval stated in the Contract Data from the starting date until Completion of the whole of the works'.* The Contract Data states the time period within which the programme has to be revised by the contractor. This may be every 4 weeks, 2 months or even longer, and is usually dependent on the complexity and overall duration of the project.

However, there is a fourth reason for a contractor to submit a revised programme for acceptance, namely:

4. *'If a compensation event has affected the Accepted Programme'.* The contractual procedure is that where a compensation event has, or will have, the effect of changing the *Accepted Programme*, then the contractor has to submit a revised programme with his quotation for the compensation event, showing the effect of the event on the *planned Completion* and *Completion Date*. However, if the contractor does not submit a revised programme with his compensation event quotation, then the project manager will make his own assessment of the effect on the *Accepted Programme* of the said event. It is suggested that upon a compensation event being implemented, then the contractor should submit a revised programme, based on the assessment of the event, to the project manager for acceptance as the *Accepted Programme.*

3.4.3 Progress updating

Lack of a formal progress updating procedure can cause failure because, without it, problems and delays will not be recognised until too late.

Even a 'perfect' programme becomes outdated unless it is updated on a regular basis. On most projects, programmes are updated monthly, but it is not uncommon to update programmes weekly or even daily.

Some of the most significant reasons for updating a project programme are to:

- Record progress
- Provide a plan for remaining work to be completed
- Provide a forecast for completion of the project and contract milestones
- Provide progress status for the project team
- Comply with contract requirements.

Maintaining accurate project records through a project control system is an important aspect of updating the programme. Information for the 'update' can come from recording progress data whilst walking the job-site, site diaries and the like kept by project supervisory staff, and status reports from subcontractors.

An important part of a progress update submission is a narrative. As well as saying what delayed progress during the reporting period, a good narrative should also explain any revisions that have been made to the programme. If a contractor states the revisions contemporaneously then the contract administrator cannot complain at a later date that he was not aware of changes to the contractor's plan.

From a contract administrator's perspective it is important to obtain a copy of the updated planning files on disk. Important because, not only are you saving a tree!, but how do you know the contractor hasn't made a mistake and accidentally forgotten to include in the 'paper' reports some of the critical or embarrassing activities. Or, for that matter, are you really going to spot a change in a logic link to an activity in a paper report? If you were an auditor, would you accept handwritten summaries of the month's transactions or would you want to see the real books?

3.4.4 Programme revisions

Regular revisions of a programme are important because the initial baseline programme is merely a plan with regards to what needs to be accomplished in order to achieve completion of the project on time. How the project actually reaches completion will most likely vary greatly from the original baseline programme, which is why regular programme revisions are crucial.

A revised programme not only records progress at the time of the revision, but should also review and, if necessary, introduce activity logic revisions to reflect current intent. These logic revisions may result in changes to the original baseline critical path.

3.4.5 Detailed review of a progress update or revised programme

A progress update or revised programme submission for review should consist of a full copy of the computer files necessary to recreate the programme, and not just the paper printouts and listings.

In addition to the five basic tests for a programme submittal listed in Chapter 8, the following checks should also be carried out on a progress update or revised programme submittal.

Check 1 System check: Most of the recognised planning software packages allow the user to determine the CPM calculation rules. For example, total float in Primavera software can be set to be computed using one of three different formulas; and mismatched dual activity predecessor links can automatically stretch out activity durations and the project completion date.

Check 2 Activity check: This involves three sub-checks:

(i) Missing 'status' information. Most planning software packages allow you to status an activity without supplying actual start and finish dates. Whilst the lack of actual dates will not affect the progress update calculations as long as the percentage progress achieved is correctly recorded against the activity, actual start and finish dates provide a good record of work already accomplished.

(ii) Deleted activities: In progress update submissions, activities that are finished should not be deleted as they form a record of when the work was achieved. However, in a revised programme submission it is acceptable that the programme include only works, and their detailed programme activities, still to be carried out on the project. One last word on deleted activities. You should never re-use the activity ID from the deleted activity as an activity ID for a new, added activity. This not only confuses the 'checker' software, it makes statistic-keeping and forensic investigation of the project very difficult. When you delete activities, also retire the activity ID.

(ii) Added activities: Adding activities should be encouraged if they are done in a way to communicate the change in the work plan. It does no good for the contract administrator or employer to insist on 'sticking to the baseline programme', i.e. updating or statusing the activities as they are completed even though the work is no longer being packaged in the manner that was originally planned.

Very little useful information can be obtained from actual start and finish dates if the activity did not describe the way the work was accomplished. Activities by their nature imply that work was being prosecuted continuously. If the work no longer proceeds in the manner envisioned, the starts and stops of work within an ill-defined work activity will make that activity no more informative than a hammock activity.

(iii) Modified activities. If an activity was neither deleted nor added, it still may have been modified. It is not 'wrong' that activities are modified. After all, the employer expects progress to be made and that involves modification of activities. The key for the reviewer of the programme is to note those modifications to activities that are other than expected progress. After spotting these types of modifications, the next step is to analyse the modifications.

Check 3 Actual dates: Any modification to an existing actual date should be accompanied by an explanation for the change. The obvious reason for this is that there is only one 'correct' date. The contractor earlier reported that the first actual date was correct. Now he or she is revising that certification. Or are they? If you fail to unambiguously affirm which is the 'correct' date, the original one or the new one, then in the event of a delay or extension of time submission, the contractor can claim that either of the dates is the correct one. Was the first date correct and a new one inadvertently entered? In other words, which of the two dates works best in the contractor's favour?

In addition to modified actual dates, you should also look for newly-added actual dates that do not fall within the update period. You should not accept new dates that just happen to fall in the future. You would think that the planning software program would prevent this from occurring, but it happens surprisingly often. Much more subtle are newly-added actual dates that fall before the start of the last progress update period. The previously- reviewed progress update showed this activity was incomplete. Now you are looking at a progress update that says you reviewed the wrong programme the previous time. Has the critical path for previous progress update moved?

Check 4 Network logic and activity links:
  (i) Where the predecessor and successor activities still exist in both the current update and previous progress update but the activity link is new, it is assumed that the contractor intended to add this relationship. These will have to be reviewed and traced individually.
  (ii) Similarly, for deleted logic links, where the predecessor and successor activities still exist in both the current update and previous progress update, it is assumed that the contractor intended to delete this relationship. These will have to be reviewed and traced individually.
  (iii) Modified logic links. For extensively modified activity links, the ramifications of changing an existing logic relationship from one type to another type is very difficult to predict without looking at each change on a case-by-case basis.

  Modifying activity link 'lags' usually results in the programme being 'stretched' or 'shortened' in a way that is very difficult to notice. This is especially true if a lot of small changes have been made to several activity link 'lags'. Many small changes can add up to one large change. You will only note this trend if you list all of these changes together in one list.

Check 5 Activity constraints: These are invisible on a plotted network, and unless you check the activity database you will not see them at all. They are very powerful and override the logic of the CPM network. Quite simply put, one constraint can completely revise an entire CRM programme.

Constraints are usually start or finish dates imposed on an activity such as 'start no earlier than' or 'start no later than'. These are more acceptable than other date constraints such as 'mandatory start'.

A careful check has to be made of the data associated with each activity to identify the constraints and a more detailed review to identify their purpose.

Remember, good programme reviews don't just happen, they take a lot of work.

3.5 *Progress records and other record keeping*

If one looks at the health of the project and compares it to the health of a vehicle, the comparisons are striking. People who ignore routine maintenance of a vehicle typically experience premature breakdowns and exorbitant repair costs that could be traced directly back to the lack of maintenance. If a contractor ignores routine maintenance of a project by taking the easy approach of updating schedules, the outcome is very likely to be an expensive 'repair' in the form of a claims battle and often a claims loss, or even missed opportunities.

Most construction professionals do not enjoy reporting progress. This task rivals the other bane of keeping minutes of meetings. That the progress-reporting duty is taken on with no enthusiasm and usually because no-one else will touch it with a bargepole, is evident in the tosh that often passes for the monthly Client Progress Report.

These reports contain more than just progress, of course. There are the usual sections in there – safety, risk, commercial, etc. – but this chapter concerns the programme/progress section. Quite often the programme and/or progress sections fall into one of two approaches:

1. The 'I'm going to prove to everyone, especially my boss, how clever I am, with lots of technical jargon and long words' approach,

or

2. The 'Let's take last month's report and just change the figures' approach.

The first approach will be almost impenetrable and unfathomable to anyone reading it, including the boss. The second approach is plain boring and is effectively saying to the client that you can't be bothered and the monthly report is unimportant.

On most projects, the client is looking for simplicity in the monthly report, and he is primarily interested in one key thing: when will the project be complete. The information in the programme/progress section of the report to the client should be easy to understand and well annotated and explained.

The format of progress reporting should be agreed with the client at the onset of the project. The programme, which will normally be maintained as a critical path network in proprietary planning software, should be capable of being summarised to level 1 barchart format.

The most readily understood graphic is that of the 'staggered-line'. Graphics of the original baseline programme and the current revised or working programme should have a vertical line showing the progress cut-off date. Progress may be indicated either by colouring along the bar or by the vertical line diverting to the actual progress position for each bar. This is a very simplistic 'Progress Indicator' chart (see Figure 5.1).

Unfortunately, often the client is left to interpret the chart for himself, and if he does not have the information to do this meaningfully he may easily jump to the wrong conclusion. Therefore, both the chart and the accompanying narrative should contain an explanation of why activities are shown in delay, what the implications are for completion of the project, and how the contractor intends to redress the situation and by when.

The simple 'Progress Indicator' chart and an accompanying narrative may be enough for many projects. However, each project is unique and there will be many where auxiliary methods are necessary, or even required under the contract. Details of three such methods are given below.

The first of these is the 'Planned Progress' chart. This addresses the volume of work, and simply measures the volume of progress in terms of activity weeks, giving no allowance for weighting of activities. Planned progress can be shown in terms of a cumulative S-curve of activity weeks achieved if the early dates are met. Another curve can be generated from the late dates. When plotted on the same chart, the area between the curves represents the zone within which the actual achievement line should lie (see Figure 3.9).

Figures are calculated after each progress update and the actual line plotted. The closer this line is to the early (left-hand) line, then the more comfortable all parties should feel. A drift towards the late (right-hand) line means that float is being used up and more activities are becoming critical.

Even though there is no weighting factor, the fact that every programme activity is taken into account means that the law of averages comes into play, and the outcome is virtually identical to one where complex weightings based on earned value or work content have been laboriously applied.

A second method is the 'Progress Tracking' chart. This is a simple but effective way of showing progress in terms of quantity or value or work done at any point. It is basically two charts in one. The x-axis is a common time scale. The left-hand y-axis shows unit per time unit (week or month) in histogram form, whereas the right-hand side relates to the cumulative figure and is shown as a simple line. The actual performance is input on a regular basis and compared to the plan (see Figure 3.10).

Because this method relies on the work being measured in the same units throughout, it is well suited for package works or individual operations or trades. It would usually be introduced to show close control of a particular critical or near-critical activity. As the planned figures are likely to be based on early dates, it is important to stress that the plan is target-based and that moderate slippage does not necessarily mean that the programme has been compromised.

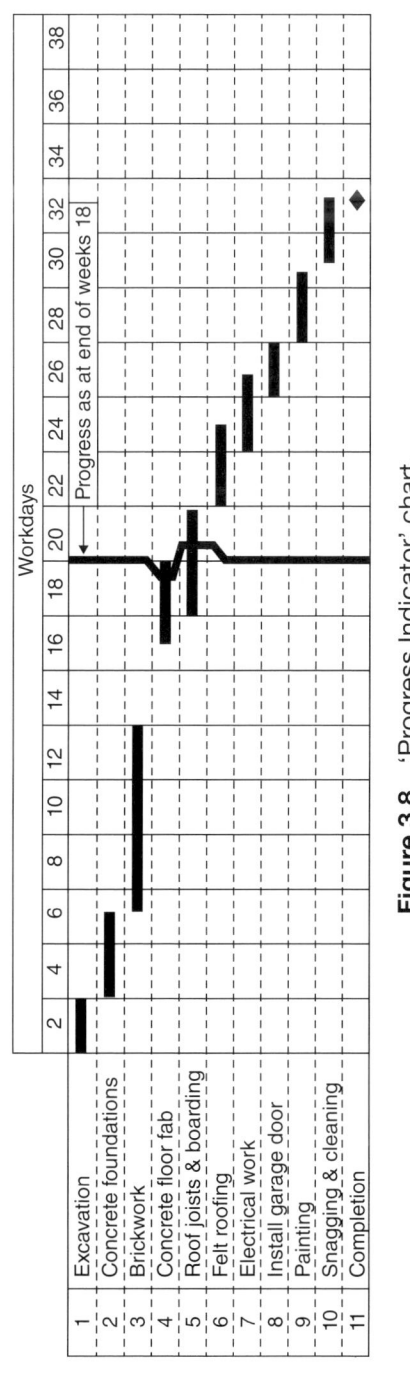

Figure 3.8 'Progress Indicator' chart.

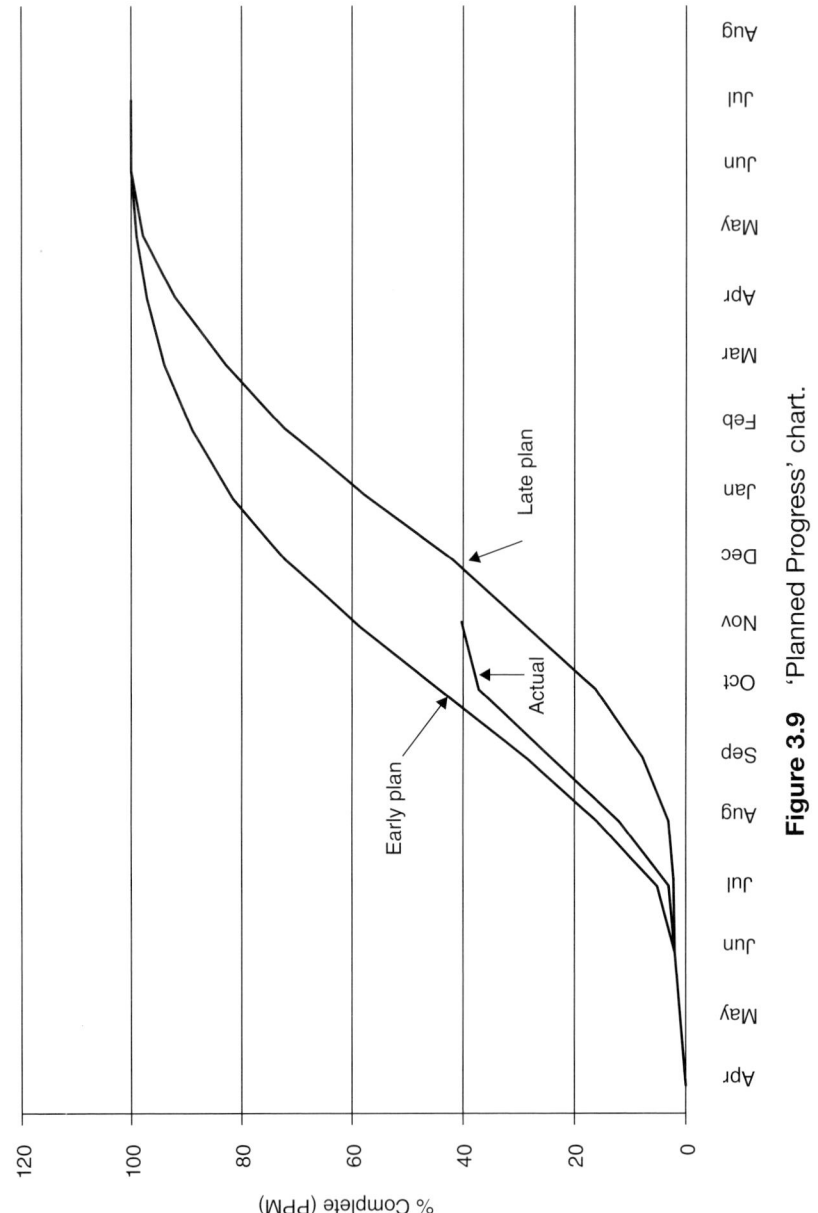

Figure 3.9 'Planned Progress' chart.

Planning, Programmes and Record Keeping

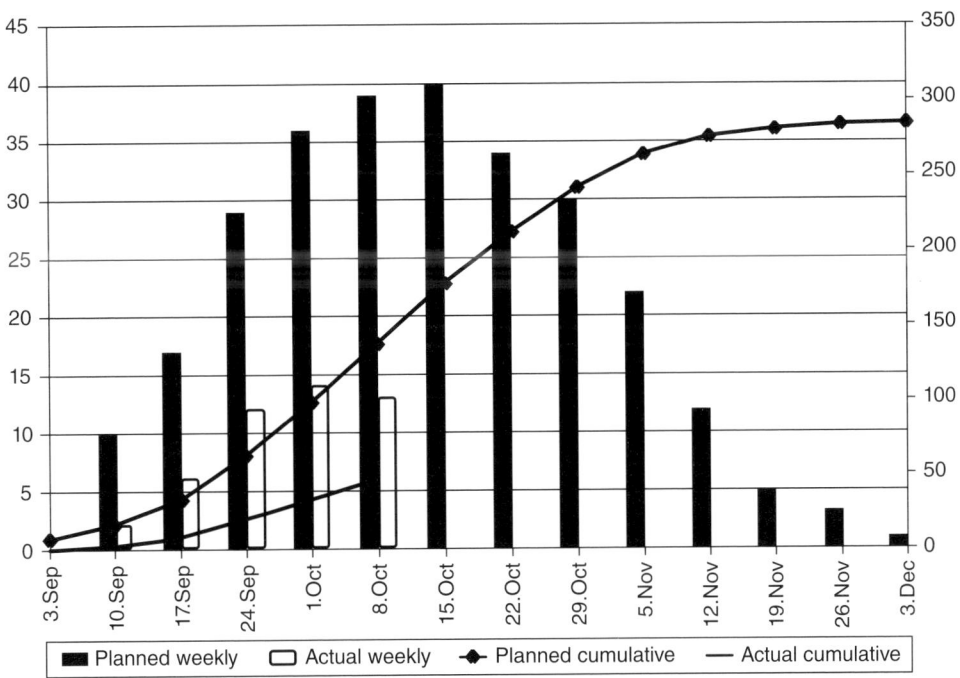

Figure 3.10 'Progress Tracking' chart.

The third method is a 'Line of Balance' chart, which comprises a series of cumulative line graphs set against a common time scale. This approach is somewhat specialist in nature and is ideal for situations of repetition, such as housing and high-rise buildings (see Figure 3.11).

The angle of each line represents the rate of output, and the gap between the lines shows the working float between operations or trades. In a situation where the lines represent recorded progress on site, it is easy to see who is delaying whom. This method also allows simple 'what-if' scenarios to be explored.

A further method of recording progress, which should be encouraged, is colouring in drawings as work proceeds. Colouring in drawings is not particularly useful when comparing progress to a plan, but it is an accessible way of showing how the site is proceeding and should not be dismissed on the grounds of crudity. Often it is exactly what is needed to convey a sense of momentum, and this method of recording progress is particularly useful in a claim situation.

However, for progress reports, this method should only cover one or two activities at a time, and one should avoid confusing the message through overkill. This method is particularly suited to activities such as piling, pipe caps, slabs, roof coverings and ceilings.

To summarise, simplicity is the watchword. Firstly, the contractor should state when the project is forecast to be complete. Secondly, the client should be

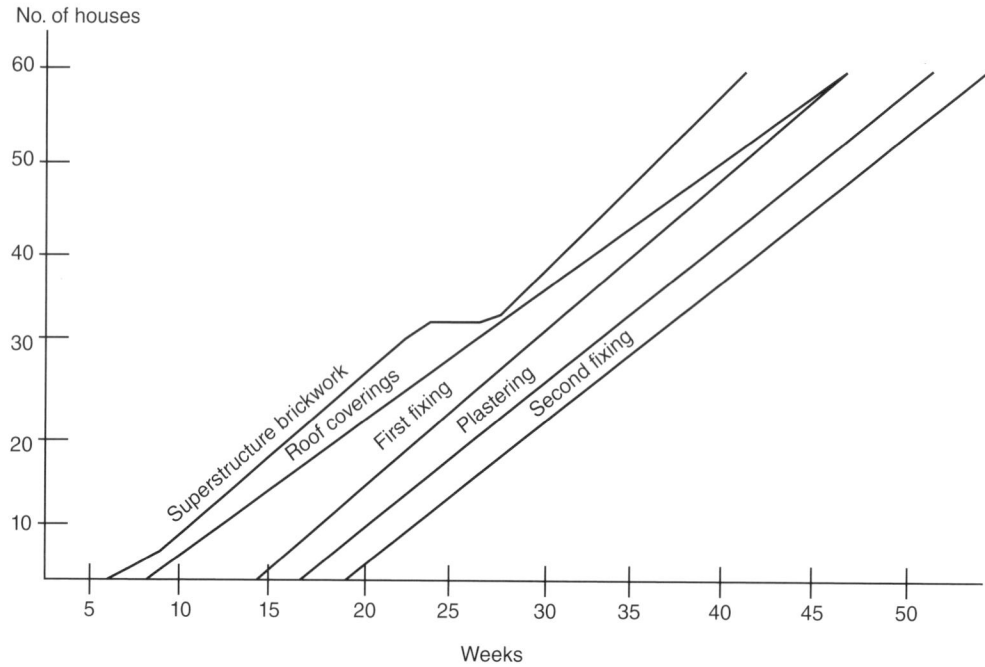

Figure 3.11 'Line of Balance' chart.

given a programme with a staggered line and the main features explained to him, including what is to be done to recoup lost time. Finally, any of the auxiliary methods, as appropriate, can be used to focus on critical areas or to get key points over.

Progress reporting should not be a complicated progress. Clients do not scrutinise boring or complicated documents to find the hidden message. It is in everyone's interest to present information in as simple a format as possible. The client will never complain of condescension if the message is clear, and if he wants something a little more sophisticated, he will ask for it.

It may be necessary in claim situations to develop a daily specific as-built programme to justify a delay claim. An as-built record of the accepted project programme will usually only record the actual start and finish dates for each activity. However, in order to support a delay and/or disruption claim it is beneficial to identify and record multiple starts and stops of affected activities.

Site diaries maintained by job-site supervisory staff are important source documents used to develop a daily specific as-built programme. The site diaries should record the following information on a daily basis:

- weather
- manpower by number and trade and subcontractor

- specific work performed, with reference to the corresponding programme activity number
- delays/interruptions/issues encountered
- work stoppages
- variation, or change order, work performed
- repair or re-work performed
- RFIs (requests for information) submitted
- CVIs (confirmation of verbal instructions) received.

A quality as-built programme can be generated from well-kept site diaries. The as-built data can be maintained in an electronic database, and nowadays can be collected using handheld devices during job-site walkthroughs. Photographs are also a very helpful way of documenting site progress; however, to be useful they should be labelled with the date when the photo was taken and with a specific description of the subject.

3.5.1 Records in a claim situation

Obtaining both a correct extension of time and time-related costs is all to do with the strength of your case, and whether you can prove it with factual records. It is a truism to say there is no substitute for good record keeping.

For example, can you prove that an entitlement to an extension of time resulted from information being received late? The key element in most cases is the contemporaneous project records. Success with the claim is all about keeping them, and then using them to demonstrate cause, effect and entitlement.

Firstly, let's deal with the programme; if you don't keep details of what was built and when, then there is very little chance of proving cause and effect. Earlier chapters explain the importance of planning and programming.

Now for the contemporaneous project records, and particularly those that relate to the work activities on the programme. Frequently, these records are just a list of percentages for the programme activities as at the date of the site meeting.

Ideally, for each activity the following progress information should be maintained for each programme activity on a fortnightly or monthly basis as a minimum:

1. As well as recording the percentage of work achieved, record the actual start and finish dates, together with any periods of inactivity or stoppages.
2. Where possible, subdivide the programme activity into smaller elements. For example, for brickwork & blockwork subdivide into elevations; for suspended ceilings subdivide into floor areas or groups of rooms.
3. Maintain allocation sheets of who did what, where and when.

4. Keep a record of the weather conditions.
5. Rcord any problems encountered together with any steps taken to overcome them including resources used.

In addition, all operational supervisory staff should keep meaningful site diaries, supplemented with photographs as these often prove very useful later if a dispute materialises. Photographs must be properly identified and the following information should be recorded on the reverse:

- The date and time when taken
- Position from where the photograph was taken
- Full details of the subject matter
- A reference number for the photograph
- Name of the photographer.

Disruption and loss of productivity are difficult to prove as generally there is very little contemporaneous data available from site showing the levels of productivity attained before and after the disruptive event. Any data that can establish disruption and productivity loss, particularly in respect of subcontractors who carry out the majority of the work, will be invaluable.

Chapter 4
Delay, Disruption and Causation

In October 2002, the Society of Construction Law published its 'Delay & Disruption Protocol'. This Protocol provides guidance to people dealing with submissions for extension of time and delay claims, both during a contract and after completion of the works. The Protocol runs to some 82 pages and was drafted by a group of experts from all sections of the construction industry.

The Protocol envisages that decision-takers, e.g. contract administrators, adjudicators, dispute review boards, arbitrators, judges, may find it helpful in dealing with time-related issues.

4.1 Delay

In this section I highlight and expand upon what I consider are four broad groups of delays, including how they operate contractually. The four groups are:

1. Non-excusable delays
2. Excusable non-compensable delays
3. Excusable compensable delays
4. Concurrent delays.

In this section, I cover the more specific aspects of:

(a) **What is delay analysis**: Delay analysis is a forensic investigation into the events or issues that caused a project to run late.
(b) **Delay analysis methodology**: I look at the two main types of delay analysis methodology, namely, *prospective* and *retrospective.*
(c) **Time and money**: The SCL's Protocol emphasises that "Entitlement to an EOT does not automatically lead to entitlement to *compensation*", and that "prolongation compensation will be recoverable if the Contractor can prove that its losses result from the Employer delay. Proper analysis of the facts may reveal the true cause without argument".

(d) **The facts**: Point (c) leads me to the fourth key aspect of this chapter: 'the facts'. The nub of any investigation and report on project delays are, or should be, the facts.
(e) **Cause and effect**: The presentation of a delay analysis is not sufficient in itself to justify compensation.

Delays happen in most construction projects, whether simple or complex. In construction, delay could be defined as the time overrun either beyond the contract date or beyond the extended date of completion. Construction project delays have a debilitating effect on the parties to a contract in terms of a growth in adversarial relationships, distrust, litigation, arbitration, cashflow problems, and a general feeling of apprehension towards each other.

Delays caused by the employer, such as late issue of drawings and specifications, frequent change orders, and incorrect/inadequate site information, generate claims from both the main contractor and subcontractors which often entail lengthy court battles with huge financial repercussions. Delays caused by contractors can generally be attributed to poor managerial skills, lack of planning and a poor understanding of financial principles, all of which have led to the downfall of many a contractor.

Because of its overriding importance for both the employer (in terms of performance) and the contractor (in terms of money), time the source of frequent disputes and claims leading to arbitration and litigation. To control this situation, a contract is formulated to identify potential delay situations in advance and to define and fix obligations to preclude such controversies. A substantial number of General Conditions clauses address this subject in one way or another. Under some circumstances, a contractor may be entitled to claim prolongation costs and loss and expense if he finishes later than the early completion schedule accepted by his employer but is still ahead of the official contract completion date. This may occur if the contractor establishes a direct cause-and-effect relationship between the employer's breach of a contractual obligation and the delay. In addition, the contractor has the burden of establishing its increased costs as a result of the delay. It is found in practice that not everything in the contract can be taken at face value and applied in cookbook fashion. Circumstances play a great part in determining which clause(s) will be applied to a particular delay claim. Also, contract law encompasses concepts of reasonableness and fair dealing, implied obligations and warranties, constructive acceleration, etc. A good general understanding of the principles involved and the operation of the applicable clauses is essential to help make appropriate decisions and take the proper action in those delay situations. In a large and complex project there will be a certain amount of give-and-take policy among the parties competing for the same time and space. Time, energy and money must not be diverted in pursuing claims and disputes over minor delays, disruptions and interferences. Accordingly, the contract conditions in some contracts contain clauses that the contractor on notice can plan for certain events. Further delays that occur arising from these events can be termed as 'non-excusable delays'.

Often contract clauses will require that a delay claim be submitted in writing within a stipulated number of days from the commencement of a delay. Further, within those stipulated number of days, after the termination of any such delay, the contractor is required to submit a written notice specifying the actual duration of the delay. Failure to give either of the above notices shall provide sufficient ground for denial of an extension of time. By giving notice, the contractor warns the employer to take alternative action to avoid or reduce the time-related costs.

However, delays do not always result from a single catastrophic event. They frequently develop slowly during the course of work. Minor delays are generally overlooked until their cumulative effect becomes financially apparent. By the time a contractor recognises that there is a problem, many different parties and natural forces may have contributed to the situation. Failure to comply with the notice requirements can contribute to the situation which may or may not defeat the claim. To avoid acceleration claims from contractors in delay situations, it is best to;

1. Issue formal extensions of time when justified.
2. Respond in a timely manner to any Notice of Claim from the contractor.

4.1.1 Types of delays

Delays can be grouped in the following four broad categories according to how they operate contractually:

1. Non-excusable delays
2. Excusable non-compensable delays
3. Excusable compensable delays
4. Concurrent delays.

4.1.1.1 Non-excusable delays

Non-excusable delays are delays which the contractor either causes or assumes the risk for. These delays might be the result of underestimates of productivity, inadequate planning or mismanagement, construction mistakes, equipment breakdown, staffing problems, or mere bad luck. Such delays are the contractor's responsibility and no relief is allowed. These delays are within the control of the contractor or are foreseeable; however, it is not necessary that they be both.

4.1.1.2 Excusable non-compensable delays

When a delay is caused by factors that are not foreseeable, beyond the contractor's reasonable control and not attributable to the contractor's fault or negligence, it may be 'excusable'. This term has the implied meaning that neither party is at fault under the terms of the contract and have agreed to share the risk and consequences when excusable events occur. The contractor will not receive compensation for the cost of delay, but he will be entitled to an additional time to complete his work and is relieved from any contractually-imposed liquidated damages for the period of delay.

4.1.1.3 Excusable compensable delays

In addition to the compensable delays that result from contract changes by instructions, there are compensable delays that can arise in other ways. Such compensable delays are excusable delays, suspensions, or interruptions.

In addition to the compensable delays that result from contract changes by instructions, there are compensable delays that can arise in other ways. Such compensable delays are excusable delays, suspensions, or interruptions to all or part of the work caused by an act or failure to act by the employer resulting from an employer's breach of an obligation, stated or implied, in the contract. If the delay is compensable, then the contractor is entitled not only to an extension of time but also to an adjustment for any increase in costs caused by the delay. Most contracts specifically address some potential compensable delays and provide equitable adjustments. The usual equitable adjustment clauses in these contracts that apply to delay are:

- changes
- differing site conditions
- suspension.

4.1.1.4 Concurrent delays

A question that frequently arises is the method of dealing with extensions of time which may be due to either or both of two causes, i.e. concurrency. The more complex the project, the more likely that this issue will arise.

Concurrent delays refer to delay situations when two or more delays, regardless of the type, occur at the same time or overlap to some degree, and either of which would have affected the project completion date.

It is important to differentiate between the delaying event or cause and the delay itself. It is generally recognised that there are times when delays may be the result of different causes but that sometimes the causes will run at the same time or overlap. This makes it difficult to decide how to treat the delay, particularly if the causes originate from different parties or the delays are of different kinds. For example, under most forms of contract, some causes may give the contractor entitlement to an extension of time; some causes may give the contractor entitlement to an extension of time and also loss and expense, whilst other causes may not entitle the contractor to any extension of time or loss and expense whatsoever.

When analysing concurrent delays, each delay should be assessed separately and its impact on other activities and the project date for completion calculated. Much will turn on the quality of planning and programming, and record keeping. Not only will there often be several delay events running in parallel, but there may be parallel critical paths to contend with and periods of acceleration and/or mitigation to take into account. The contract conditions will also have to be taken into account on the analysis technique used.

4.1.2 Analysis of project delay

Over the past 20 years the review of project delay has become more analytical and a more challenging feature of construction law. Is this overshadowing the factual evidence?

As Mr Justice Dyson noted in the Henry Boot Construction v. Malmaison Hotel (Manchester) case:

> "It seems to me that it is a question of fact in any case, as to whether a relevant event has caused, or is likely to cause, delay to the works beyond the completion date."

In the late 1980s I became involved in the investigation and review of delays to construction projects. At that time, this task was the domain of quantity surveyors. However, with the development of user-friendly project planning software, 'delay analysis' became the province of those who, like myself, came from a planning background.

4.1.2.1 What is a 'delay analysis'?

Delay analysis is a forensic investigation into the events or issues that caused a project to run late. Delay analysts refer to 'critical' and 'non-critical' delays; the first are events causing delay to the project's completion date and the second type affect progress on the project but do not directly impact the project completion date.

During the past decade, developments in computer technology and the availability of more advanced planning software packages have, in my view, changed the way in which delay claims and the results of a delay analysis are presented.

4.1.2.2 Delay analysis methodology

I do not want to dwell here on the types of delay analysis (e.g. 'as-planned impacted', 'as-built but for', 'time-impact', etc.), but to look at the two types of delay analysis methodology.

The first type of delay analysis methodology is *prospective*. This demonstrates the theoretical or likely impact of the consequences of delaying events rather than showing what in fact occurred. The basis of this methodology is to establish a programming model of the project, usually the contractor's as planned programme, then impact the model by the application of delaying events. This type of methodology is commonly used to demonstrate what extension of time a contractor is due as a result of the application of employer responsible delaying events. This is said to be the contractor's *entitlement*. Entitlement in this context is derived from the results of a delay analysis and is not to be confused with contractual entitlement. In summary, the prospective type of methodology is a theoretical calculation of the likely delay a delaying event(s) would cause to project completion. In other words, it focuses firstly on the delaying event and then demonstrates the likely delay to progress and ultimately project completion that is likely to flow from the event.

The second type of delay analysis methodology is *retrospective*. The retrospective analysis tries to show what actually occurred on a project, where the delays were, and what caused the delay to project completion. The analysis shows how actual progress differed from what was planned. By focusing on how the works actually progressed, the analysis will show when work activities were delayed and, from the results of the analysis, investigation of what caused the actual delays can be carried out. In summation, this type of methodology looks at what actually happened, what activities were actually delayed and only thereafter what caused the delay.

Both types of delay analysis methodology are to some degree subjective. The prospective analysis relies heavily on a programming model of the project and the delay analyst's opinion on how the delay event was likely to influence the model. The retrospective analysis is, in my opinion, less subjective as it relies on actual progress. However, interpretation of the results as to what caused delay is subjective. This is because the delay analyst will usually have to consider a number of related issues as to what caused the delay and apply his own experience and judgement.

The Society of Construction Law's Delay and Disruption Protocol provides guidance, procedures and mechanisms to manage change on a project. The Protocol recognises the two types of methodologies and it has separate sections that address each type.

4.1.2.3 Time and money

My explanation and views on the two types of delay analysis methodologies are primarily related to extensions of time. But in many instances parties and experts use the results of a delay analysis to establish the effect of delay in terms of money. This is, in my view, unsound and incorrect. Indeed, in relation to compensation for prolongation, the SCL's Protocol emphasises that: "Entitlement to an EOT does not automatically lead to entitlement to *compensation*", and that "prolongation compensation will be recoverable if the Contractor can prove that its losses result from the Employer Delay. Proper analysis of the facts may reveal the true cause without argument".

One of the arguments that often arises is 'as to the time when recoverable prolongation is to be assessed; is it to be assessed by reference to the period when the employer delay occurred or by reference to the extended period at the end of the contract'. The Protocol's answer to this question is that: "Once it is established that compensation for prolongation is due, the evaluation of the sum due is made by reference to the period when the effect of the Employer Risk Event was felt, not by reference to the extended period at the end of the project."

4.1.2.4 The facts

"Now, what I want is, Facts." You may recognise this quotation from Charles Dickens' novel *Hard Times*. The nub of any investigation and report on project delays are, or should be, the facts. By this I'm talking about the project's factual records, i.e. variations (and their like), correspondence, minutes of meetings, diaries, progress reports, etc. These are the 'facts'. However, too often do we see extension of time and delay claim submissions containing several lever-arch file of these 'facts', with no specific linkage to the alleged events that caused delay. There may also be a bundle of computer printouts indicating the claimed effect – but the causal link is not clearly defined. The referee (judge, arbitrator, or contract administrator) is expected to find it; and often it's like looking for a needle in a haystack! The primary purpose of the delay claim submission is to assist the referee to weigh up the provisions of the contract, relevant case law, witness evidence, contemporary factual records, as well as considering the results of a delay analysis to form his own view.

4.1.2.5 Cause and effect

As I said earlier, the presentation of a delay analysis is not sufficient in itself to justify compensation. It is necessary to establish:

(a) **The event**: the event to be identified as a fact, e.g. late supply of information, to a contractor.
(b) **Liability**: determined by interpretation of the contract.
(c) **Effect**: the change to the planned progress of the works as a result of the event. This may be demonstrated by a prospective delay analysis (for entitlement to an extension of time), and/or a retrospective delay analysis to

assist in compensation. Sometimes the contract provisions may determine the methodology of delay analysis that is required, i.e. estimated future delay and/or the probable future delay (prospective analysis), or the actual delay (retrospective analysis).

(d) **Compensation**: the actual costs incurred as a result of the effect of the event. The SCL's Protocol gives clear guidance on this issue: "*Compensation for prolongation should not be paid for anything other than work actually done, time actually taken up or loss and/or expense actually suffered. In other words, the compensation for prolongation caused other than by variations is based on the actual additional cost incurred by the Contractor.*"

(e) **Causation**: The causal connection between the event, effect and compensation. In some instances the identified causative event may have been caused by a previous causative event. For example, delay caused by winter working may have been caused by the project being delayed into winter due to an earlier causative event. Therefore the chain of causation and the incidence of any secondary causative events will need to be investigated and established.

4.1.3 Conclusion

This brings us back to the question posed earlier; does an over-reliance on theoretical calculation of delay overshadow reliance on factual records?

In my opinion, yes. From my experience as an expert on time-related matters and my adjudication experience, I have come across many instances where a party, and/or its expert, has put forward a brief report supported by a mass of computer-generated barcharts and very little linkage to contemporaneous factual information. The idea seems to be, "That's what the computer says – so it must be right!" Well, we have all heard of the saying 'garbage in, garbage out', haven't we. Delay claims and results of investigations into project delays are being presented on a 'what would have occurred' basis, using theoretical models, rather than on a 'what did occur' basis, and interrogation of the facts. Facts are the best means of persuasion.

Don't get me wrong, I consider that a delay analysis is important as *part* of an investigation into project delay – but it does not provide the complete answer. A credible delay analysis should (a) sit comfortably with the party's presentations (pleadings, etc.), witness evidence, and contemporaneous information, and (b) provide results to be incorporated into a 'cause and effect' matrix.

Finally, I believe that the SL's Delay and Disruption Protocol will encourage the use of techniques that will produce credible delay analyses. The 'ground rules' as set out in the Protocol will, if followed, crystallise delay analysis methodologies and give credence to a well-presented delay analysis in a time-related dispute. But the delay analysis should be a supporting document as part of an interrogation of the facts.

4.2 Disruption

Claims for disruption and additional costs are routinely made during the course of a project yet they remain notoriously difficult to prove.

One of the main reasons for this is that productivity losses are often difficult to identify and distinguish at the time they arise. This is different than other money claims which are more directly concerned with the occurrence of a distinct and compensable event together with a distinct and direct consequence, such as an instruction for a discrete variation during the progress of the works or a properly notified compensation event.

As such, most claims for disruption are dealt with retrospectively. The claimant is forced to rely on contemporary records to try and establish a causal nexus for identified losses (cause and effect) which are all too often inadequate for the purposes of sufficiently evidencing a loss of productivity claim. When this happens the claimant is often forced into the situation where it advances a weak global or total cost claim of sorts to try and recover some of its losses.

The cause-and-effect burden of proof is the same for a claim for loss of productivity as for any other claim insofar as the claiming party must first establish that the event or factor causing the disruption is a compensable risk event under the contract. To do this, the contract needs to be reviewed to understand the basis of the agreement as certain productivity issues may have been foreseeable and therefore accounted for within the claimant's productivity allowances. The contract may also identify if a party expressly accepted certain productivity risks.

Where courts and tribunals have a clear focus on linking cause and effect, claims for disruption will come under greater scrutiny. It is unlikely that contractors and subcontractors will succeed where their claims for disruption are based simply on a global overspend on labour or plant for the whole of the contract working period. Sufficient detail is required to isolate the cause of the disruption complained of and evaluate the effects of that disruption.

In October 2002, the Society of Construction Law published its 'Delay & Disruption Protocol'. This Protocol provides guidance to people dealing with submissions for extension of time and delay claims, both during a contract and after completion of the works. The Protocol runs to some 82 pages and was drafted by a group of experts from all sections of the construction industry.

The Protocol envisages that decision-takers, e.g. contract administrators, adjudicators, dispute review boards, arbitrators, judges, may find it helpful in dealing with time-related issues.

4.2.1 So what is 'disruption'?

Disruption (as distinct from delay) is disturbance, hindrance or interruption to a contractor's normal working methods, resulting in lower efficiency. If caused by the employer, it may give rise to a right to compensation either under the contract or as a breach of contract.

Phrases commonly used to define disruption include ripple effects, knock-on impacts, secondary effects, impact on unchanged work, lost productivity. Typically, the direct impacts of changes can be documented or estimated reliably even if entitlement issues are argued. However, the real secondary effects of changes, i.e. disruption costs, are problematic.

Productivity losses occur from disruption whether or not it occurs on the critical path and causes delay to completion. Loss of productivity means that the work progresses more slowly and, as is often the case, float is consumed or at least this is alleged to be the case.

Disruption is also concerned with more nebulous matters such as short interruptions, difficult working conditions, working around others and plant and materials being in the wrong place. It is the overall loss of productivity from these sorts of events that can also cause overall delay if the disrupted activity is on the critical path.

Disruption and loss of productivity are difficult to prove as generally there is very little contemporaneous data available from site showing the levels of productivity attained before and after the disruptive event. Any data kept that can establish disruption and productivity loss, particularly in respect of subcontractors who carry out the majority of the work, will be invaluable.

While courts and tribunals have accepted the idea of disruption costs, cases attempting to claim disruption generally have not been very successful. Historically, it is the disruption component of claims that has been the most difficult to quantify, the most contentious, and has resulted in the lowest recovery for claimants.

In the nature of things, disruption losses are harder to demonstrate than delay. There are two main reasons for this.

First, disruption claims are often based on myriad facts often occurring at the same time that are difficult to marshal at all let alone disentangle and analyse. In addition, they are often poorly recorded.

Second, labour productivity is a matter of human reaction to a situation and is inherently variable. What the productivity would have been compared to what it actually was, and why, is almost automatically a matter for debate.

Nevertheless, however difficult disruption claims may be they are of the utmost importance.

The losses caused by even minor disruptions can be considerable. Academic studies have shown that disruptions lasting longer than half an hour cause productivity loss of between 20% and 40% over a day depending on the trade.

The degree of success of a disruption claim is likely to be directly proportional to the quality of the supporting records and documentation. Good records also give a much better chance of recovery of disruption costs during the course of the works as well as in any subsequent dispute resolution arena.

It is essential for the contractor to maintain and make available to the contract administrator good site records in order that the contract administrator may carry out proper assessments of disruption. The contractor should also give prompt notice of disruption whether it is required by the contract or not. The contract administrator can then promptly investigate the claim and he and the employer will be in a much more difficult position later if he does not.

4.2.2 Why are disruption claims so difficult?

Disruption can be widely separated in space and time from the causative event(s), but to be claimed successfully must be causally tied to their source. Disruption impacts can be cumulative across large numbers of individual impacts. Disruption is fundamentally about productivity, which is hard to measure and thus rarely measured well.

The ideal form of damage quantification is to define the amount of impact, including disruption, that would put the injured contractor in the condition it would have been but for the damaging events – a challenging analytical task. Disruption claims must screen out the effects of other concurrently occurring contributors, such as strikes.

Contractor/employer discussions of project cost growth tend to be adversarial even while the project continues, making efforts to quantify, explain and mitigate disruption especially challenging.

Finally, with all these difficulties, there is also, quite frankly, a poor track record of rigour in disruption quantification. It is far easier for both sides to put all blame on the other without rigorous analysis to back their claims. Assertions, such as "you caused all our problems", and the counter-assertion "you mismanaged everything" are common.

4.2.3 Essentials of a disruption claim

Three elements must be proven for a disruption claim to be successful.

1. **Liability**: This is a legal issue, which is separate from quantification of impact.
2. **Damage**: The resultant damage is the claimed loss of productivity. In essence, it is an equitable adjustment of the difference between what it would have reasonably cost to perform the work as originally required and what it reasonably cost to perform the work as changed. This may be relatively

straightforward for small numbers of isolated changes to a project. In cumulative impact claims, the challenge is how to consider all changes together with everything else that may have been happening on a project.

3. **Causation**: This requires the establishment of a causal link between the direct impacts claimed and the total injury suffered by the claimant. This is the most difficult element to prove of the three, because of the complexity of separating inefficiencies that are caused by the employer's changes from those that would have occurred anyway, for example, as a result of contractor-caused inefficiency.

Disruption is fundamentally about reduced productivity and increased rework on the project. To quantify disruption rigorously and accurately, a methodology needs to be able (a) to explain the variations in productivity, and (b) to assess what would have happened under alternate conditions. But to quantify the disruption accurately is not enough; to be successful in resolving disputes, a methodology must also address the other challenges to resolving a disruption claim. It must be able to:

- Provide a causal map that cogently ties a resulting effect to the precipitating event, even though portions of the effect may be weeks or months downstream and in a different part of the project.
- Account for and explain the synergy among the individual claimed events that can result in a much larger than expected overall impact.
- Explain why productivity trajectories would have been affected, how much and for how long.
- Account for other concurrent events that might also influence project performance, such as strikes, difficult labour markets or mismanagement.

Ideally, the methodology would be auditable and different assumptions could easily be tried, enabling it to serve as an objective test bed. Additionally, the methodology should permit validation.

4.2.4 Contemporaneous project records

Maintaining contemporaneous project records are important for both the contractor and the employer.

A daily log of what has happened and why is the best kind of record. The ideas is to have daily reports from all the foremen on a site about all the work that they have completed, the disruptions to their gang's work and the perceived causes.

By concentrating site labour productivity on the incidences of disruption, the resultant labour productivity levels can be used objectively and systematically to quantify the effects of disruption on site productivity.

The more information about site conditions and how the labour force was affected, the better. As much as practicable, a contractor should be able to demonstrate the plan of what was to be done with what resources and then shows the events that meant the plan could not be followed and what had to be done instead.

Other contemporaneous project records which should be maintained and logged are:

- **Requests for Information (RFIs) and Technical Queries (TQs)**. These will be important and should record when the information is required. They will be even more powerful if they warn of what the consequences are likely to be if the information is not forthcoming as required.
- **Written records of all instructions**. If the employer or his contract administrator issues instructions orally then the contractor should always confirm these in writing. If the likely effect can also be identified then that will be an advantage.
- **Photographs**. These are always helpful in establishing a disruption claim if it is difficult to convey in written form the real effect of, for example, a congested working area and photographs can convey the impact. Photographs are more compelling if they are taken regularly from the same vantage points. Comparisons can be made with work that proceeded productively and there is no suggestion of the angles being carefully chosen to make things look worse than they were. Videos are also worth considering, particularly for congested working. However, a photographic record is open to interpretation on its own and it should not be seen as a substitute for proper written records.

All these contemporaneous records still have to be used; and the best way is to use them as a source to build a narrative presentation to show that the complaints that are being made had an effect on site. The causes being relied on should therefore be correlated with the perceived causes shown in the records

The narrated effects, however, still do not mean much on their own until they are measured in some way and that is very difficult to do. The method of doing this can contribute to presenting a convincing case on cause and effect.

The key way in which it can do so is by eliminating effects that are not the responsibility of the employer. However, in some sense all of the approaches to quantifying the effects of disruption have an element of the global about them. It is impracticable, even with perfect records, to present a case as to the quantified effect of every individual event of disruption. What has to be found is the best method of making a sound assessment that reliably measures the effects of the complaints when they are grouped together.

4.2.5 The SCL Protocol

In October 2002, the Society of Construction Law published its 'Delay & Disruption Protocol'. This Protocol provides guidance to people dealing with submissions for extension of time and delay claims, both during a contract and after completion of the works. The Protocol runs to some 82 pages and was drafted by a group of experts from all sections of the construction industry.

The Protocol envisages that decision-takers, e.g. contract administrators, adjudicators, dispute review boards, arbitrators, judges, may find it helpful in dealing with time-related issues.

There are 21 'core statements of principle' in the Protocol. Of these, only one statement directly concerns disruption:

> 21. "Disruption; disruption (as distinct from delay) is disturbance, hindrance or interruption to a Contractor's normal working methods, resulting in lower efficiency. If caused by the Employer, it may give rise to a right to compensation either under the contract or as a breach of contract."

However, there are three other core statements which are somewhat related to disruption, acceleration, and/or mitigation. These are:

> "1. Programme and records; to reduce the number of disputes relating to delay, the Contractor should prepare and the Contract Administrator (CA) should accept a properly prepared programme showing the manner and sequence in which the Contractor plans to carry out the works. The programme should be updated to record actual progress and any extensions of time (EOTs) granted. If this is done, then the programme can be used as a tool for managing change, determining EOTs and periods of time for which compensation may be due. Contracting parties should also reach a clear agreement on the type of records that should be kept.
>
> 13. Mitigation of delay and mitigation of loss; the Contractor has a general duty to mitigate the effect on its works of Employer Risk Events. Subject to express contract wording or agreement to the contrary, the duty to mitigate does not extend to requiring the Contractor to add extra resources or to work outside its planned working hours. The Contractor's duty to mitigate its loss has two aspects – first, the Contractor must take reasonable steps to minimise its loss; and secondly, the Contractor must not take unreasonable steps that increase its loss.
>
> 20. Acceleration; where the contract provides for acceleration, payment for the acceleration should be based on the terms of the contract. Where the contract does not provide for acceleration but the Contractor and the Employer agree that accelerative measures should be undertaken, the basis of payment should be agreed before the acceleration is commenced. It is not recommended that a claim for so-called constructive acceleration be made. Instead, prior to any acceleration measures, steps should be taken by either party to have the dispute or difference about entitlement to EOT resolved in accordance with the dispute resolution procedures applicable to the contract."

4.2.5.1 Observations

My observations on the above Core Principles,

1. **Core principle 13: Mitigation of delay and mitigation of loss**. A clear exposition of the situation. More could have been said in the Protocol about the contractor's rights, or otherwise, to claim reasonable costs of mitigation.
2. **Core principle 20: Acceleration**. This is a broadly correct interpretation of the position, but the reference to the possibility of accelerating by instructions about hours of working and sequence of working is to be doubted.
3. **Core principle 21: Disruption**. The definition of disruption does not adequately explain that disruption can also refer to a delay to an individual activity not on the critical path where there is no resultant delay to the date for completion. The Protocol also states that most standard forms do not expressly deal with disruption; that, of course, is true. However, the JCT forms refer to regular progress being materially affected. That appears to be broad enough to encompass both disruption and prolongation.

The Protocol's 'Guidance Section 2' deals with guidelines on preparing and maintaining programmes and records. However, there is not a great deal of guidance on maintaining records generally.

Stress is placed on obtaining an 'Accepted Programme'; that is, a programme agreed by all parties. There are several problems with this. Perhaps the foremost is that the architect will be unlikely to have the requisite skills and/or experience or, indeed, the information required to accept the contractor's programme. He is probably capable of questioning parts of it, but highly unlikely to be possessed of sufficient information to be able to satisfy himself that the programme is workable. The Protocol, rightly, accepts that the contractor is entitled to construct the building in whatever manner and sequence he pleases, subject to any sectional completion or other constraints. The Protocol states:

> "Acceptance by the CA (contract administrator) merely constitutes an acknowledgement by the CA that the Accepted Programme represents a contractually compliant, realistic and achievable depiction of the Contractor's intended sequence and timing of construction of the works."

This is placing a responsibility on the architect (or CA as the Protocol prefers) which he is not required to carry. There appears to be no need for a programme to be accepted. It is sufficient if the contractor puts it forward as the programme to which he intends to work. The architect is entitled to question any part which appears to be clearly wrong or unworkable. But, in the light of the contractor's insistence that he can and will carry out the works in accordance with the submitted programme, it is difficult to refuse a programme unless firm objections can be raised.

The Protocol also recommends that the Accepted Programme be updated with progress at intervals of one month, and more frequently on complex projects. The Protocol describes the updating process as follows:

> "Using the agreed project planning software, the Contractor should enter the actual progress on the Accepted Programme as it proceeds with the works, to create the Updated Programme. Actual progress should be recorded by means of actual start and actual finish dates for activities, together with percentage completion of currently incomplete activities and/or the extent of remaining activity durations. Any periods of suspension of an activity should be noted in the Updated Programme. The monthly updates should be archived as separate electronic files and the saved monthly versions of the Updated Programme should be copied electronically to the CA, along with a report describing all modifications made to activity durations or logic of the programme. The purpose of saving monthly versions of the programme is to provide good contemporaneous evidence of what happened on the project, in case of dispute."

All of this is good and sensible advice.

4.2.6 Quantification methods

Two of the quantification methods more commonly used are the following.

4.2.6.1 The measured mile

The most appropriate way to establish disruption is to apply a technique known as 'the measured mile', which compares the productivity achieved on an unimpacted part of the contract with that achieved on the impacted part.

The great advantage of this method is that the base figures come from the performance of the contractor on the project with the comparison being made on the man-hours expended or the units of work performed. In theory, this should eliminate disputes about under-tendering because of optimistic productivity assumptions, unrealistic programmes, and inefficient working or other matters that may have affected productivity that are not the responsibility of the employer.

To make this method truly effective and persuasive it needs to have the following features:

- The work on the measured mile should be similar in type, nature and complexity to the disrupted work.
- The numbers skill level and supervision of the labour needs to be similar.
- The productivity levels demonstrated by the measured mile will need to look reasonable otherwise credibility will be damaged.

- The difference between the actual productivity of the affected work and the normal productivity on the measured mile results solely from the events being complained about.
- The measured mile needs to allow for the contractor's risks and inefficiencies.
- The measured mile needs to be a big enough sample size to give confidence that it is a valid comparison.

If difficulties arise in finding an unimpacted part of the project, then comparison of productivity on other contracts carried out by the contractor is the next best method. This is acceptable provided that sufficient records are available from the other contracts, and providing the comparison is on a like-for-like basis.

4.2.6.2 Earned value

This method uses the standard outputs or norms, i.e. that it takes a certain number of hours to complete a particular task, which are then used in estimating.

At the end of each day, the tasks completed therefore represent a number of hours that have been earned. If those 'earned hours' are divided by the actual hours spent on labour, then the resulting ratio is a measure of productivity for the activity concerned. In theory, this measure is needed for every disrupted activity, but in practice it may be sufficient to use a number of key or characteristic activities or to treat them in groups. For example, looking at all formwork or rebar fixing without distinguishing the different types or in M&E installation there is sometimes a particular item that has to be installed for every room or area whatever other variations in size or type of installation.

The method is entirely dependent on accurate information as to exactly what work has been completed when and the availability and accuracy of relevant norms used. If there are good records then a presentation of changing earned value ratios combined with a correlated chronology of disruptive events and the reports of the perceived causes of disruption on site can be the basis for a convincing case.

However, it will be necessary then to link the detailed site information and productivity loss to the actual causes of disruption that are relied on. One major problem with the earned value method is that it does not allow for low productivity caused by the contractor as a result of its own inefficiencies or otherwise. If allowance can be made for these, then the presentation will be more likely to succeed. A contractor will say that is in the norms, but the question will remain as to performance on the project concerned.

It is worth noting that both the measured mile and earned value ratios will show a much greater loss of productivity than by simply adding up lost time shown by records. The difference can be dramatic. There have been cases where the records are good and show around 5% of hours to have been lost but the loss of productivity is in the region of 30%.

4.2.6.3 Other methods of quantification

If a measured mile is not available from the project or a comparable project carried out by the contractor, and information does not exist to permit earned value to be used, then the next best approach can be to use 'model productivity curves and factors'. These have been developed by a number of organisations from data collected on a range of protects, such as The US Army Corps of Engineers, The Mechanical Contractors Association of America, and The Chartered Institute of Building.

Various industry organisations have developed models and factors that can be used to estimate productivity losses. Productivity loss factors assist in matters such as overtime working and over-manning. If the relevant fact situations can be shown to be assessed then the use of this kind of general statistical data may be presented as a reasonable measure of the effects. It does have the advantage that it can be used prospectively and individual types of effect can be isolated if the appropriate statistics are available.

However, this approach will always be vulnerable to the charge that it is not grounded in project information and so is to a degree hypothetical.

Another method is the '*global claim*' or '*total cost claim*'. A global claim is a claim for the difference between actual cost and the contracted cost. The cause of the difference is said to be all the matters that have caused the disruption. In its simplest form there is no attempt at any explanation as to how the matters complained of caused the additional costs above those that were expected There may be a global claim for the whole of the works or for some part of it. That method of itself means that it ignores other paternal causes of the losses such as those caused by the contractor itself. In large measure, claims presented like this rest on an inference that the complaints caused the loss.

However, where records are poor and/or the disrupting events were extremely complex and/or were project-wide, it may not be possible to present a claim in any other way. If this is the situation then a global claim is unlikely to be found objectionable as such. The principle of global claims has been upheld in the common law jurisdictions

In the UK, it has been held that, where it is impossible or impractical to separate findings for each matter or break down the complex interaction of events, then the global or composite award may be allowed, as long as there is no element of profit or double-counting.

Indeed, it can be argued that if the breach and loss are shown then causation may be presumed if there is no reason to believe that the two are unconnected. However, that may do no more than shift the burden of proof to the employer to show other causes of disruption.

There are serious risks attached to global claims because, as set out above, to a greater or lesser degree tribunals require claims for compensation to be proved.

The tests may be more or less strict but always exist in some form. There is therefore a risk to the contractor and an opportunity for the employer that the claim will fall because the events that are the responsibility of the employer have not been shown to have brought about the loss that has been suffered.

The claim will be more vulnerable to be attacked by the employer showing other causes of the loss of productivity that has been suffered for which he not responsible than are the other methods referred to above. Then there is a high risk of the whole global claim collapsing.

A global claim will be less risky if it is modified (*modified global claim*) so as to adjust the lender or estimate to take account of any underbidding or any matters that the Contractor acknowledges the more detailed this exercise is the better modified global claim). The project is, for example, sometimes rebid retrospectively to work out what it would have cost the contractor without the matters for which the employer is responsible.

The risks of the global claim are also reduced if it is broken down to some degree. So if it is possible to quantify some issues in isolation then it should be done. If parts of the project can be broken down into a number of mini-global claims for groups of activities and only the overall balance is the main global claim, this will be an improvement. There will be a better chance of part of the claim succeeding and the contractor will show that he has done what he can to demonstrate causation, thereby giving greater justification for the global approach to the balance of the claim.

Another way of diminishing risk is to put forward material in the global claim that would enable its apportionment if the tribunal were to find that it was justifiable in part, but that it had failed to take account of other matters that were not the responsibility of the employer.

There may be legal issues as to the ability of a tribunal to apportion but it is certainly an arguable approach even under the Common Law.

Percentage additions is a last resort where the contractor is unable to do more than show that there was culpable delay by the employer and cannot present any reasoned case as to causation or quantification. This approach is very vulnerable to attack but may be worth advancing, especially if the percentage addition sought is relatively modest and there is a good case that there must have been at least some loss. The approach may also enjoy more sympathy from a tribunal on smaller projects where the resources for recordkeeping and sophisticated analysis may not exist.

Clearly not all disruption attracts the payment of compensation. The contractor may be entitled to compensation for the effects of lost productivity to the extent that a breach of obligation exists, a causal link to the offending party can be established and the effects of that disruption calculated appropriately. Most standard forms of contract do not deal expressly with disruption. If they do not, then disruption may be claimed as being a breach of the term generally implied into construction contracts, namely that the employer will not prevent or hinder the contractor in the execution of its work.

4.3 Causation

The overall concept of causation is easy to grasp: a contractor is only entitled to additional time and money insofar as that additional time and money has been caused by something which entitles him to time or money as the case may be.

First, there is a fundamental question as to the meaning of 'caused'. In some contexts, the courts have applied a 'but for' test, i.e. 'A' could be said to have been caused by 'X' if it would not have happened but for 'X'.

Keating on Building Contracts contains an interesting passage on this issue. The hypothetical case is: A contractor claims payment under the contract, e.g. for delay resulting from variation instructions and there is a competing cause of delay. The competing cause of delay could be:

(a) no one's fault, e.g. bad weather, or
(b) the contractor's own delay in breach of contract.

Four possible approaches are considered:

1. **The Devlin approach**. If a breach of contract is one of two causes of a loss, both causes cooperating and both of approximately equal efficiency, the breach is sufficient to carry judgement for the loss;
2. **The dominant cause approach**. If there are two causes, one the contractual responsibility of the defendant and the other the contractual responsibility of the plaintiff, the plaintiff succeeds if he establishes that the cause for which the defendant is responsible is the effective, dominant cause. Which cause is dominant is a question of fact, which is not solved by the mere point of order in time, but it has to be decided by applying commonsense standards;
3. **The burden of proof approach**. If part of the damage is shown to be due to a breach of contract by the plaintiff, the claimant must show how much of the damage is caused otherwise than by his breach of contract, failing which he can recover nominal damages only.
4. **The tortuous solution**, i.e. the application of the 'but for' test.

Whilst the general rule is that causation must be proved by showing on the balance of probabilities an effective causal connection between the breach of contract or tort and damages, the courts have shown a degree of flexibility in their approach.

However, a party wishing to prove causation in cases where there are complex facts has to ensure that its case is clear and easily understandable by both by the other side and the tribunal. In the difficult area of proof of causation, properly prepared computer-generated analyses can be useful provided that they are founded in reality and are seen as a means to prove a case rather than proof in themselves.

Whatever evidence is presented, there must be a clear understanding of what it is meant to establish, and the evidence must be in a form which can be readily understood by the tribunal. Whilst a fundamentally poor case cannot be improved upon by the method of presenting evidence, there are many good arguable claims where proper presentation of the evidence can assist in establishing the case.

4.3.1 Presenting a successful claim

I now consider how a successful claim can be presented.

The first requirement is a proper analysis of the facts. There is a natural tendency to assume that the task is impossible and therefore to avoid any proper analysis. Some causative events are bound to be better than others in the sense that they are stronger in terms of liability and more likely to lead to an inference that some substantial degree of delay or cost is likely to have been caused by that event. The analysis of the facts should establish not only the scope and extent of the event but also what the immediate effect was.

For example, an instruction to change the type of radiators to that originally specified as a bare allegation is unlikely to persuade a tribunal but if someone can explain the scope and extent of the necessary design, supply and installation activities and explain what immediate effect it had on the productivity of the works, then that will enable a party to show a tribunal that the foundation for an inference is present.

The claim should be divided up into discrete parts. This may be a division based on particular areas of a project, particular activities or trades or particular time periods. It may be a combination of these. This allows the case to be presented to the tribunal in a logical manner and assists the tribunal to understand the factual position more easily. A global claim which piles up allegations and evidence in the hope that the tribunal will be overwhelmed by the complexity and will accede to a large claim is likely to fail outright.

The use of a sophisticated computer-generated analysis which conceals everything except an end result is a dangerous approach. First, the tribunal has no way of understanding the evidence. Second, all analyses are based on assumptions and if a major assumption is shown to be wrong the analysis will be worthless. If it is more transparent, for instance by considering the sensitivity of the analysis to the assumptions, it is likely to overcome this difficulty. Third, the facts which have been analysed must be apparent and explainable for the same reason.

Computer-generated analyses are, however, frequently useful provided that certain steps are taken. It is important to have a baseline 'as planned' and an actual 'as-built' programme. These should be the subject of early expert discussions in an attempt to reach agreement or narrow the differences between the parties. Any analysis performed on a baseline programme must be firmly rooted in reality. It must take into account any changes in resources or in logic which have been made. It should have a narrative so that the change can be explained in terms of cause and effect. It must, of course, also be factually correct. The purpose of such an analysis is to show that there is a nexus between cause and effect, or at least that there is an inference to be drawn linking the two.

Often parties seek to present a number of different approaches to try and support the overall conclusion that they should succeed. In my experience, the greater the number of different approaches that are used in a given case, the less the impact of the chosen or primary analysis. Frequently, the different approaches can be shown to have inconsistencies which undermine all of the approaches. Whilst a party may choose to consider a number of ways of analysing the position, it is important that a decision is taken on the single approach which is used to present the case. Where it is thought that different approaches might strengthen the case, it is necessary to confirm that such a conclusion is correct.

Whatever analysis is produced, it is important to present the key findings of the analysis in a clear way. It is not just the overall conclusion which is of importance but the steps along the way. Therefore, in order to prove a delay an acceleration and/or loss of productivity claim a party must show that, on the balance of probabilities, the circumstances complained of arose by reason of matters which entitle him to compensation. That has been refined into the premise that he must prove not only the events themselves but that they actually caused the acceleration and/or loss of productivity. In other words, cause and effect, i.e. causation.

4.3.2 First cause or ultimately critical

A related question concerns supervening events. What is the position where a delay is first caused by one matter, but then is subsequently or partly overtaken by a delay arising from another cause. In other words, should a retrospective delay analysis operate on a first cause basis or an ultimately critical basis.

The choice between *first cause* and *ultimately critical* is probably the most important unanswered causation question among experts in this field. It may be helpful to start with a non-building example.

Suppose man A needs to cross a desert. In order to survive the journey he will need water for the trip. He has two enemies, B and C, both of whom wish to kill him. Both take action shortly before A's departure. B poisons the water in A's can. C, unaware of B's action, makes a small puncture in A's can. A duly sets off; when he comes to drink his water, he finds it has gone and he dies.

The question is: Who caused A's death? The method *first cause* will have it that B caused A's death. Whereas, the method called *ultimately critical* will have it that C caused A's death.

The *first cause* in this scenario would run as follows. By poisoning A's water, B took a step that was sufficient of itself to ensure A's death.

This is essentially the approach that was taken by the House of Lords in the gruesome case of *Baker v. Willoughby* [1970] AC 467 where the facts were as follows.

The plaintiff was injured in the leg as a result of a motor accident caused by the defendant's negligent driving. That injury caused the plaintiff pain and suffering, and adversely affected the plaintiff's earning capacity. The plaintiff was then involved in an attempted robbery, during which he was shot in the same leg in such a manner as to require amputation of that leg.

The question, of course, was whether the defendant was liable to the plaintiff for a lifetime's worth of pain and suffering and loss of earnings, or just the pain and suffering that occurred prior to the amputation.

The House of Lords said that the defendant was liable for the full lifetime. Lord Reid said that the defendant did not have his liability reduced if, as he put it, "The later injuries merely become a concurrent cause of the disabilities suffered by the injury inflicted by the Defendant".

There are some important features of *first cause*:

(a) It is necessary to start the analysis from the base of a contract programme, in order that there may be a calculation of what would have been the effect of that delaying event alone if there had not been any subsequent delaying events.
(b) If there are successive causes of delay (as there always are in these complex cases) then the analyst is logically bound to construct a new contract programme in respect of the position following every delaying event. In these programmes, he must suppress his knowledge of subsequent delaying events.
(c) It is thus evident that *first cause* carries with it a subjective element, or at any rate a high degree of professional judgement. It may well be incumbent on the contractor, following the first cause of delay, to look at the ways in which the delay might be mitigated. This is particularly important when one comes to consider resourcing issues. Suppose, for example, a bricklaying team starts a 10-day job repointing a wall. They plan to move from west to east. On the third day they encounter an obstruction, in the form of an old advertising board which should have been removed by others but which has not yet been removed. Do they stop in their tracks and wait for the advertising board to be removed? Of course not, they continue work at the other end of the wall and little or no time is lost.

4.3.2.1 Ultimately critical

Ultimately critical will tell you that the man in the desert above was killed by enemy C, i.e. one has to look at what was ultimately the cause of A's death, taking advantage of the benefit of hindsight. A died, not of poisoning, but of thirst.

This is the approach that was adopted in *Jobling v. Associated Dairies* [1982] AC 794. The facts here were as follows.

The plaintiff suffered an accident at work which left him with a back injury and a consequent reduction in earning capacity. A few years later, but before the trial of his action against his employers took place, he was found to be suffering from myelopathy, a condition with which the accident had no connection but which rendered him very soon after its discovery totally unfit for work.

The House of Lords held that there could be no recovery for loss of earnings from the time of total incapacity; the myelopathy was not to be disregarded since the court must recognise that the supervening illness would have overtaken the plaintiff in any event.

The House of Lords thought that the Baker v. Willoughby case may have been wrongly decided.

It should be clear that, under the ultimately critical method, the sequence planned within the original contract programme is largely irrelevant. What matters is what actually happened. To this extent, ultimately critical is much more straightforward and more objective than first cause.

There is however a complication. In the case of first cause the analyst is, in effect, keeping track of the delay, rather than of the project duration. In the case of ultimately critical, on the other hand, the analyst's first task is to identify the string of critical activities which in total add up to the actual project duration. Some of those activities will be characterised as delay, and some of those activities will be characterised as contract work which was always necessary. How does the analyst separate the wheat from the chaff?

Suppose, for example, a contract period is 50 weeks, but the project in fact takes 60 weeks to complete. Using a first cause the analyst will look at the first cause of delay, to which he might ascribe, say, one week. He will then look at the next cause of delay, to which he might ascribe another week of critical delay, bringing the total of critical at that point to two weeks. By the time he has finished analysing the project he will have to ensure that the total number of critical weeks' delay he has analysed adds up to the actual delay, i.e. 10 weeks.

Conversely, under ultimately critical, the analyst's first task is to identify what was in fact the critical path through the actual 60 weeks of the project. He must then identify how much of that period he ascribes to delay, and where.

In answer to the 'how much' question, he obviously looks at the contract period. Here, the contract period was 50 weeks, and accordingly the amount of time ascribed to be as delay is 10 weeks.

The question of where he puts that 10 weeks is much more difficult. In practice, most analysts would usually pay great attention to the contract programme here.

Chapter 5
Loss of Productivity

5.1 Introduction

One of the most contentious areas in construction and claims is the calculation or estimation of loss of efficiency, or lost productivity. Unlike direct costs, the loss of efficiency is often not tracked or cannot be discerned separately and contemporaneously. As a result, both causation and entitlement concerning the recovery of disruption costs are difficult to establish. Compounding this situation, there is no uniform agreement within the construction industry as to a preferred method of calculating the loss of efficiency. There are, in fact, numerous ways to calculate loss of efficiency and the resulting productivity factor.

In Section 5.2, I highlight and expand upon what I consider are three of the most important aspects of loss of productivity.

5.1.1 Productivity and efficiency

Productivity loss is experienced when a contractor is not accomplishing its anticipated achievable or planned rate of production. It is best described as a contractor producing less than its planned output per work hour of input. Thus, the contractor is expending more effort per unit of production than originally planned.

Productivity and efficiency are critically important in the context of construction and engineering contracts, both large and small. Construction contractors are typically paid for work completed in place that conforms to the terms of the contract. This is sometimes referred to as pay item work and is generally true whether the contract is lump sum/firm fixed price, cost reimbursable, target cost, unit cost or pay item work or as a percentage of previously-defined categories of work, often referred to as a bill of quantities.

Practical Guide to Disruption and Productivity Loss on Construction and Engineering Projects,
First Edition. Roger Gibson.
© 2015 John Wiley & Sons, Ltd. Published 2015 by John Wiley & Sons, Ltd.

5.1.2 Common causes of loss of efficiency

Later in this chapter I describe 24 of the most common causes of loss of efficiency. I also give my detailed recommendations that the contractor should:

(a) reconfirm the baseline estimate to ascertain that the project estimate is basically correct. In doing so, the contractor can ensure that the efficiency loss detected is not simply the result of comparing onsite productivity and efficiency to a flawed baseline.
(b) gather all necessary supporting documentation and file the lost productivity, or efficiency, claim as prescribed in the contract.

5.1.3 Methods of productivity seasurement

In this section I describe the two most common methods of productivity measurement:

(a) The measured mile approach, and,
(b) The earned value approach.

Both these methods are discussed in detail, together with their respective pro's and con's.

5.1.4 Worked example

In the latter part of this chapter, I give details of a Worked Example, using the earned value approach. I have based this on an actual submission document which has been redacted.

I now describe in detail the three important aspects of loss of productivity.

5.2 Productivity and efficiency

Whether it is in manufacturing, white-collar work, professional sports, or construction, productivity is one of the major components that distinguish success from failure. In construction, productivity has become even more important as budgets and time frames are tightened to the point of strangulation. As a result, the ability to measure productivity and to articulate deviations in productivity has become essential for the success of businesses involved in the construction and engineering industries.

Simply stated, productivity is a measurement of rate of output per unit of time or effort usually measured in labour hours. For example, cubic meters of concrete placed, linear meters of conduit or pipe installed per hour.

Productivity loss, therefore, is experienced when a contractor is not accomplishing its anticipated achievable or planned rate of production; it is best described as a contractor producing less than its planned output per work hour of input. Thus, the contractor is expending more effort per unit of production than originally planned. Therefore, a challenging aspect of construction cost control is measuring and tracking work hours and production in sufficient detail to allow analysis of the data in order to determine the root cause(s) of poor labour productivity, i.e. efficiency.

Productivity and efficiency are critically important in the context of construction and engineering contracts, both large and small. Construction contractors are typically paid for work completed in place that conforms to the terms of the contract. This is sometimes referred to as pay item work and is generally true whether the contract is lump sum/firm fixed price, cost reimbursable, target cost, unit cost or pay item work or as a percentage of previously-defined categories of work, often referred to as a bill of quantities.

Therefore, unlike companies such as car manufacturers, construction contractors are rarely paid on the basis of the entire completed product. Furthermore, for contractors, productivity is related to project cash flow and profitability.

All too often in construction, the terms 'productivity' and 'production' are used interchangeably. This is, however, incorrect. Production is the measure of output (i.e. things produced) whereas productivity is the measurement of the production.

Given this set of operating terms, it is therefore possible for a contractor to achieve 100% of its planned production but not achieve its planned productivity. For example, a contractor could well be accomplishing the planned rate of production of 300 linear metres of pipe per day in the ground but be expending twice the amount of labour planned to accomplish this daily production rate. In this case, the contractor would be accomplishing 100% of planned production but operating at 50% productivity.

Thus, production and productivity are not reciprocal numbers. It does not necessarily follow that if a contractor is 75% productive then they are 25% inefficient.

Measurement and allocation of responsibility for loss of efficiency can be difficult. There are a number of reasons for this difficulty. Amongst them, are the following:

(a) Loss of efficiency resulting from some action which is the responsibility of the owner may not be easily detected or observed at the outset. Unless a contractor has a good productivity monitoring plan, all that may be known at the outset of a problem is that the onsite crews are not

completing work activities as planned, and project schedule, costs and cash flow are suffering as a result. As a result, appropriate written notice to the project owner is often not promptly issued.
(b) Efficiency is frequently not discretely tracked on construction projects in a contemporaneous manner. Unless a contractor uses some sort of structured earned value system for tracking output units and input units, there is no way to measure efficiency and productivity contemporaneously. Thus, loss of efficiency with the degree of certainty demanded by many owners can be difficult to prove.
(c) There are myriad ways to calculate loss of efficiency and productivity. There is no common agreement amongst cost professionals as to how such lost hours should be calculated. Notwithstanding this statement, there is general agreement amongst cost professionals that a comparison to unimpacted work on the project is generally preferred when there is sufficient data available.

Finally, after efficiency loss is calculated, it is still difficult to establish causation. Contractors tend to blame such losses on owners and ask to be compensated. Owners, on the other hand, often blame underestimation by the contractor or poor management and site organisation and thus deny additional compensation for lost efficiency.

An employer's typical criticism of loss of efficiency claims is that the calculation of the loss lacks certainty or precision. The irony of this criticism is that, more often than not, the very factors that cause a contractor to suffer efficiency losses are the precise reasons why detailed direct cause and effect records cannot be accurately maintained to calculate damages on a discrete impact-by-impact basis. For example, generally a contractor's project records and accounting system do not isolate separate costs for efficiency and productivity losses from other costs of the project because the work impacted by the loss is integral with base contract work.

5.2.1 Method development

During the 1970s, petroleum and chemical companies (petro-chem) understood the importance of monitoring construction labour productivity on their major construction projects. As a result, petro-chem championed a system of monitoring construction progress against labour hours expended. In order to service petro-chem requirements, engineering procurement construction (EPC) companies of that time provided cost engineers and cost-reporting systems for the project. The technique implemented to measure labour productivity was a standard man-hour system. The basis for the system utilised time studies to develop base manhours for the installation of various commodities. These time studies resulted in standard labour units for installation of commodities such as reinforcing steel, concrete, structural steel, process piping, and electrical conduit and cable. Petro-chem and EPC estimating departments expanded the standard labour units to include a multitude of labour functions such as equipment setting and alignment, welding, instrumentation and controls. The standard hours were based on the work being performed under certain conditions such as a 40-hour workweek, unobstructed access to the job site, moderate temperatures, and so on. In order to quantify the standard manhours earned on a project at a given point in time, cost engineers would perform physical surveys and calculate the quantity of commodities installed. The quantity installed was then multiplied by the standard unit man-hour factor for that commodity to develop the manhours earned. The earned manhours were then compared to the actual manhours expended installing the commodity. In most cases, the earned manhours were divided by the actual manhours, resulting in a productivity index. A productivity index of less than 1, meaning that the actual manhours were greater than the progress earned manhours, indicated the project was experiencing lower-than-anticipated productivity. A productivity greater than 1, where the progress earned manhours were greater than the actual manhours, indicated work was being performed better than anticipated.

Electrical activities such as hanging conduit, pulling wire, installing fixtures and devices, and terminations would be monitored as electrical activities. The standard unit rates used to monitor the progress were in fact usually identical to the unit rates used by the estimators, with the possible exception of an adjustment for local labour factors. The local labour multiplier is used to adjust for differences in labour conditions, work schedule and unique project complexity issues. By employing the standard manhour labour productivity method, EPCs were able to measure changes in labour productivity and exert management controls to mitigate losses. Further, a project was able to evaluate labour productivity on foundations compared with labour productivity on erecting steel or labour productivity on installing electrical components. All work activities were weighted in accordance with their standard manhour component.

During the period from 1970 to 1990, it was quite common to have up to six quantity surveyors/cost engineers providing monthly labour reports on projects in the £50 M–£80 M value range. However, economic pressures have forced EPCs, as well as others, to reduce field staffing, and in many cases eliminate cost engineers. In their place, other methods have been established for measuring labour productivity that are very similar to the original standard manhour method. As a starting point, let us examine how a typical contractor develops its lump sum price. Most contractors use their own historical standard units or sometimes refer to standards in estimating manuals such as Spons, which are then adjusted based on the contractor's historical database. Just as with EPCs, contractors perform rigorous takeoffs of design drawings to determine their quantities and multiply those quantities by the standard units. The resultant manhours are then multiplied by a rate to include labour pay, taxes, insurance, fringe benefits, and associated overhead and profit. In a similar manner, contractors use the quantity takeoff to obtain supplier quotes and unit prices for material needs and then mark them up for associated overhead and profit. The two are combined, and the overall price is submitted.

If it is deemed that the bid is the lowest and best price, the contractor is awarded the contract. Following the award, the contractor submits to the architect/engineer a price breakdown of the project for the interim payment application. The contractor develops this breakdown by determining from the estimate the amount of labour and material costs associated with work activities in a particular area of the project. In most cases, the breakdown of the project must be in sufficient detail to allow the architect/engineer and the contractor to agree on a physical percent complete based on a visual observation. The breakdown, in all cases, must be approved by the architect/engineer. It is often required that the value for labour and materials be separated in an agreed-upon breakdown. To develop this breakdown, the contractor refers to the estimate based on quantities and unit manhours.

Once the project breakdown has been agreed upon, the architect/engineer review and approve, on a monthly basis, the progress complete achieved by the contractor. The more detailed information included in the project breakdown increases the accuracy of the observed percent complete. Using this method, the architect/engineer and the contractor are agreeing upon a labour percent complete of the project on a monthly basis. This percent complete is in fact based on the estimate weighting.

On a monthly basis, the contractor will have a labour report that provides manhour expenditures on the project. The manhour expenditures can be compared to the percent of labour progress earned. It is recommended that the non-contributing labour such as nonworking general foreman or supervisory staff be removed in order to enhance the labour productivity measurement.

Another important element is to crosscheck the quantities used in the interim payment application breakdown against actual quantities in order to identify any significant variances from the estimate. Major variances should be adjusted before the calculation. Also checks should be made to determine that the estimated units are in fact proportional to normal standards. A high or low local adjustment factor will have a negligible effect on the overall productivity analysis. The productivity measurement techniques described above form the foundation for the present-day productivity comparison analysis known as the measured mile (see page XX).

5.3 Common causes of loss of efficiency

On construction projects there are numerous circumstances and events which may cause efficiency to decline. The following list of causes is, although not all-inclusive, fairly well covers the majority of situations encountered on a construction project.

The circumstances set forth below may all impact labour efficiency. However, for a contractor to successfully recover damages due to loss of efficiency from the employer, the contractor will need to clearly demonstrate that the root cause of the event or circumstance was something for which the employer was responsible.

Additionally, the contractor must be able to show a cause-and-effect relationship between the event and the impact to labour efficiency in order to recover damages (i.e. costs and/or time). However, the recoverable damages are not limited to direct costs. They may also include ripple damages or indirect costs, to the extent that a cause-and-effect relationship can be established between the downstream effects and the originating event.

(a) **Absenteeism and the missing man syndrome.** When a crew hits its productive peak the absence of any member of the crew may impact its production rate because the crew will typically be unable to accomplish the same production rate with fewer resources or, perhaps, a different mix of skill and experience levels.

(b) **Acceleration (directed or constructive).** The deliberate or unintentional speeding up of a project may result in lengthy periods of mandatory overtime, the addition of second shifts, or the addition of more labour beyond the saturation point of the site or that can be effectively managed or coordinated. All of this may have distinct impacts on productivity and loss of efficiency.

(c) **Adverse or unusually severe weather.** Some bad weather is to be expected on almost every project. But pushing weather-sensitive work from good weather periods into periods of bad weather, or encountering unusually severe weather, may impact labour efficiency.

(d) **Availability of skilled labour.** To be productive and efficient, a contractor must have sufficient skilled labour on site. To the extent that skilled labour is unavailable and a contractor is required to construct a project with less skilled labour it is probable that efficiency and productivity will be impacted.

(e) **Changes, ripple impact, cumulative impact of multiple changes and rework.** All projects encounter some change during construction. This is to be expected. Some authors believe that 5–10% cost growth due to changes is the expected norm. However, major change (change well beyond the norm), change outside the anticipated scope of work (cardinal change), multiple changes, the impact of change on unchanged work, or the cumulative impact of changes may all affect efficiency. The need to remove work already in place, the delays attendant to changes, the need to replan and resequence work, for example, may also cause efficiency to decline.

(f) **Competition for craft labour.** If a nearby project(s) commences concurrently with the start of a project that was estimated and planned to utilise a stated level of labour skill and availability, and a competition for that skilled labour base ensues, productivity may be adversely impacted. Financial incentives, work rule changes and other issues may result in labour leaving one site for another, resulting in lower productivity, loss of efficiency and increased costs for the first contactor. Further, the replacement labour may be more costly and/or less skilled.

(g) **Craft turnover.** If a crew suffers from continual craft turnover, it is unlikely that it will achieve good productivity, simply because one or more members of the crew may be on the learning curve and thus decrease the overall efficiency of the entire crew.

(h) **Crowding of labour or stacking of trades.** To achieve good efficiency, each member of a crew must have sufficient working space to perform their work without being interfered with by other craftsmen. When more labour is assigned to work in a fixed amount of space it is probable that interference may occur, thus decreasing efficiency. Additionally, when multiple trades are assigned to work in the same area, the probability of interference rises and efficiency may decline.

(i) **Defective engineering, engineering recycle and/or rework.** When drawings or specifications are erroneous, ambiguous, unclear, etc., productivity and efficiency is likely to decline because crews on site are uncertain as to what needs to be done. As a consequence, crews may slow down or pace their work, or have to stop all together while they wait for clear instruction.

(j) **Dilution of supervision.** When crews are split up to perform base scope work and changed work in multiple locations or when work is continually changed or resequenced, onsite supervisory staff are often unable to effectively perform their primary task – to see that crews work efficiently. Onsite supervision ends up spending more time planning and replanning than supervising. It is probable that efficiency will decline because the right tools, materials and equipment may not be in the right place at the right time.

(k) **Excessive overtime.** Numerous studies over many years have consistently documented the fact that productivity and efficiency typically decline as overtime work continues. The most commonly stated reasons for this result include fatigue, increased absenteeism, decreased morale, reduced supervision effectiveness, poor workmanship resulting in higher-than-normal rework, increased accidents, etc. One author has gone so far as to suggest that "…on the average, no matter how many hours a week you work, you will only achieve fifty hours of results". The premise underlying this statement is that, while overtime work will initially result in increased output, if it is continued for a prolonged period, the output may actually decline for the reasons stated earlier. Thus, long-term overtime may lead to increased costs but decreased efficiency. The effect of continued overtime work on labour productivity is perhaps one of the most studied productivity loss factors in the construction industry.

(l) **Failure to coordinate trade contractors, subcontractors and/or vendors.** If the project management team fails to get subcontractors, material or equipment to the right place at the right time, then productivity and efficiency may decline as crews will not have the necessary resources to accomplish their work, various trades interfere with others or work is not available for the crews to carry out.

(m) **Fatigue.** Craftsmen who are tired tend to slow down, make more mistakes than normal, and suffer more accidents and injuries, thus efficiency may decrease for the entire crew.

(n) **Labour relations and labour management factors.** When there are union jurisdictional issues, industrial relations issues, unsafe working conditions or other safety issues, multiple evacuation alarms in existing facilities, untimely issuance of permits, access issues, etc., labour efficiency may be adversely impacted in multiples ways.

(o) **Learning curve** Typically, at the outset of any project, there is a learning curve while the labour crews become familiar with the project, its location, the quality standards imposed, laydown area locations, etc. This is to be expected and is typically included in a contractor's estimate. However, if the work of the project is shut down for some period of time and labour crews laid off, then when work recommences the labour crews brought back to the project may have to go through another learning curve. This is probably an unanticipated impact on labour efficiency. If this happens more than once, then each time a work stoppage occurs another learning curve productivity loss impact may occur.

(p) **Material, tools and equipment shortages.** If material, tools or construction equipment are not available to a crew at the right location and time, then the crew's efficiency will probably suffer as they may be unable to proceed in an orderly, consistent manner. Similarly, if the wrong tools or improperly sized equipment are provided, efficiency may also suffer.

(q) **Overmanning.** Efficiency losses may occur when a contractor is required to or otherwise utilises more personnel than originally planned or can be effectively managed. In these situations, efficiency losses may occur because the contractor may be forced to use unproductive labour due to a shortage of skilled labour; there may be a shortage of materials, tools, or equipment to support the additional labour; or the contractor may not be able to effectively manage the labour due to a dilution of supervision.

(r) **Poor morale of craft labour.** When, for example, work is constantly changed or has to be torn out and redone, the morale (i.e. enthusiasm for their work) is likely to suffer. When this occurs, efficiency may decline.

(s) **Out of sequence work.** When work does not proceed in a logical, orderly fashion efficiency is likely to be negatively impacted, for example, as crews are moved around the site haphazardly.

(t) **Programme compression impacts on productivity.** Contractors are not legally bound to prove that contract performance was extended to recover for lost efficiency. When there are delays early on in the project, compression of the overall timeframe for later activities is often looked to as the way to make up for delays and finish the project on time. From a strict planning and programming perspective this may be possible to do without accelerating individual work activities by utilising float in the project's overall schedule. However, on many projects, programmes are not fully resource loaded. As a consequence, a properly updated programme reflecting the delays may show the project finishing on time, without shortening individual activities. It may result in overmanning of the work by the contractor due to the shortening of the overall duration allowing the contractor to complete the total remaining work. This is known as programme compression. Programme compression, when associated with overmanning, often results in significant efficiency losses due to dilution of supervision, shortages of materials, tools or equipment to support the additional labour, increased difficulty in planning and coordinating the work and shortages of skilled labour.

(u) **Rework and errors.** When work on site must be done more than once in order to get it right, efficiency may suffer as a result.

(v) **Site or work area access restrictions.** If a work site is remote, difficult to get to, or has inefficient or limited access then productivity may suffer because labour, equipment and materials may not be on site when and as needed to support efficient prosecution of the work. In addition, efficiency losses may occur when access to work areas are delayed or late and the contractor is required to do more work in a shorter period of time, which may result in overmanning, dilution of supervision and lack of coordination of the trades.

(w) **Site conditions.** Physical conditions (such as saturated soils); logistical conditions (such as low-hanging power lines); environmental conditions (such as permit requirements prohibiting construction in certain areas during certain times of the year); legal conditions (such as noise ordinances precluding work prior to 7:00 am [0700 hours] or after 6:00 pm [1800 hours]) may all negatively impact productivity and efficiency on a project.

(x) **Untimely approvals or responses.** When employers, designers and/or construction managers fail to respond to contractually-required submittals or requests for information in a timely manner, efficiency on a project may decline as crews may not have the authority or sufficient knowledge to proceed with their work.

Once the first efficiency loss has been detected, the contractor should reconfirm the baseline estimate to ascertain that the project estimate is basically correct. In doing so, the contractor can ensure that the efficiency loss detected is not simply the result of comparing onsite productivity and efficiency to a flawed baseline.

Once this effort is complete, the root cause of the efficiency loss needs to be determined. If the causation is found to be something for which the owner is liable, recommended practice is to follow the mandates of the contract with respect to providing written notice to the appropriate party as soon as possible.

Subsequently, recommended practice is to gather all necessary supporting documentation and file the lost productivity or efficiency claim as prescribed in the contract. Some contracts allow claim filing within a specified period of time after the notice of claim was filed whereas others provide for claim filing within so many days after the event or circumstance has passed. Regardless, the contractor seeking to recover lost productivity and/or efficiency costs should follow the mandates of the contract as closely as possible.

5.4 Methods of productivity measurement

Two of the most popular methods of measuring productivity are:

(a) The 'measured mile' approach
(b) the 'earned value' approach.

I now explain in detail these two methods.

5.4.1 The 'measured mile' approach

The measured mile analysis employs the productivity techniques described above to compare different periods of productivity within a project. This comparison is often used to explain and quantify the effect that different conditions have on a labour force's ability to perform. The measured mile represents the labour force's ability to perform on the particular project at hand versus a theoretical calculation.

For example, if ABC Electrical contractor obtains a project away from its normal area, a measured mile analysis will identify the base efficiency at which ABC is able to employ a different labour force. Further, the measured mile can be used to determine the labour inefficiencies caused by delay, disruption, or interference on a project. If it can be determined that there is a period of unhindered or least hindered time on a project in which the labour expended reflects an efficient use of the labour force, then a ratio can be established between physical work accomplished and actual manhours expended. This time and associated percentage of work accomplished and related actual manhours provide a ratio of manhours to percent (manhours/percent) that becomes the measured mile.

The measured mile period is then compared to the impacted period, which in turn allows for a calculation of lost time associated with the impact. Further, if the employer is responsible for the delay or disruption, the contractor may be entitled to a claim for the added labour hours associated with the inefficiency. In some projects, the impact is very clear, such as in the case of acceleration. If the first 60% of the project was proceeding at a normal 40-hour workweek for eight months and for eight months a reasonably consistent level of productivity was achieved, then if for whatever reason the employer directed the contractor to accelerate the completion of the project by working 10 hours a day, seven days a weefrom 9 April 2007 k, starting on the first day of the ninth month, the resultant loss of productivity becomes apparent.

Labour productivity would be measured subsequent to the authorisation of this acceleration and compared to the previous period. The contractor, however, has the burden of establishing a causal link between the impact caused by the employer and the contractor's increased time and cost. The measured mile approach has been found to be a reasonable approximation of those actual costs incurred.

The key advantage of a measured mile approach is that it relies on data agreed to by the architect/engineer on a contemporaneous basis during the actual contract performance. The labour productivity levels for both the measured mile and the impact periods are derived from project records, payroll records, and interim payment applications.

By employing the above crosschecks and adjustments, and having the parties work together on progress monitoring, the criticisms of using percent of labour as a measured mile are minimised. The labour percent complete is preferred over attempting to isolate a single commodity such as the installation of 1-inch conduit because there are many more construction activities associated with the success of the installation than just installing the conduit. It is important to note that it is widely recognised that the measured mile approach avoids the shortcomings of industrial studies or estimating guidelines because it is tied to the actual performance at the job site.

5.4.1.1 Support for measured mile analysis

One of the more common criticisms of a measured mile analysis is that a measured mile can only be performed on same or identical work activities. An example of such a position is that one can only measure productivity of exactly the same type of work, such as installing 1-inch branch conduit on the third floor of the building, and this cannot be compared to the labour productivity of installing2-inch feeder conduit on the fourth floor. However, in 2001, the US Department of Veterans Affairs (VA) Board of Contract Appeals ruled that a contractor could use the measured mile method of calculating labour productivity, even though it was impossible to compare identical impacted and less impacted work activities.

The contractor in this case was P.J. Dick Inc. (PJD), who was awarded a contract by the VA to construct a clinical addition to a medical centre in Ann Arbor, MI. PJD subcontracted the electrical work to Kent Electrical Services (KES), on a time and material basis. PJD alleged that the VA's electrical design was incomplete and in error, and these problems resulted in delays and inefficiencies.

The VA refused to grant a time extension and ordered Dick to accelerate. Dick's subcontractor, Kent, was forced to add crews and perform the work in an accelerated, disruptive manner.

Dick paid Kent for all the increased labour required and then pursued a claim against the VA. Dick's expert, Mr Apprill, used a measured mile analysis. Mr Apprill determined that all installation of branch circuits had been affected by design problems or acceleration. He examined other electrical work performed and determined that the installation of feeder circuits was sufficiently similar to branch circuit installation. Both branch and feeder circuits use the same basic materials of conduit and wire and were installed by union electricians. However, feeder circuits were long continuous vertical runs of large size conduit and did not involve any device installation. Mr Apprill compared the branch circuit work against productivity achieved on the installation of feeder circuits prior to the acceleration.

The VA's expert expressed a general objection to the measured mile methodology on the basis that feeder circuit work is not the same as branch work. The expert argued that the measured mile methodology requires good and bad period productivity performance of one crew performing the same work and that since the feeder and circuit work involved different crews, Dick's measured mile analysis was fatally flawed. The VA Board of Contract Appeals disagreed: "We find no basis to conclude that either the productivity of the same crew or exactly the same work is a prerequisite for a valid measured mile analysis to establish the amount loss of productivity. We agree with the GSA Board of Contract Appeals when it held in *Clark Concrete Contractors, Inc., 99-1 BCA 30,280* [1]:

> "(The government) is correct in asserting that the work performed during the periods compared by (the contractor) was not identical in each period. We would be surprised to learn that work performed in periods being compared is ever identical on a construction project, however. And it need not be; the ascertainment of damages for labor inefficiency is not susceptible to absolute exactness (citation omitted). We will accept a comparison if it is between kinds of work which are reasonably alike, such that the approximations it involves will be meaningful.
>
> On balance, we find that Mr. Apprill's approach to quantification of the VA-caused productivity loss is reasonable and valid. We recognize that feeder circuit work generally involves installation of larger sized electrical conduit and wire in longer, straighter conduit runs. However, KES' labor for feeder circuit installation was drawn from the same labor pools used for branch circuit work, and work are reasonably

similar enough to branch circuit work to permit a valid comparison. The work was performed in the working conditions planned and budgeted by KES. Consequently, we find PJD's measured mile analysis to be a reasonable approximation of the effect of the VA-caused inefficiencies under the Clark Concrete Contractors standard."

In addition to providing the above clarification that the work need not be identical, the board goes on in a quantum discussion to describe the merits of the measured mile analysis.

"We, as most other courts and boards, recognize that quantifying the loss of labour productivity is difficult and that the determination of the dollar amount of damages for labor inefficiency with exactitude is essentially impossible. In recognizing this fact, we expect that measurement of the amount of inefficiency would usually be supported by expert testimony. The use of a "measured mile" analysis developed by a qualified expert is recognized as the most reliable, though not exact, methodology to quantify labor inefficiency. Clark Concrete Contractors, Inc. GSBCA No. 14340, 99-1 BCA 30,280; W.G. Yates & Sons Construction Company, ASBCA No. 48,398, 01-2 BCA 31,428; U.S. Industries, Inc. v. Blake Construction Co., 671 F.2d. 539 (D.C. Cir. 1982); Luria Bothers & Co. v. United States, 369 F2d. 701 (Ct. Cl. 1966)."

In the Clark Concrete Contractors case the General Service Board of Contract Appeals upheld Clark's use of a measured mile analysis to quantify labour productivity decreases as a result of the government's design changes. Clark Concrete Contractors, Inc., was contracted to build an FBI field office in Washington DC. As a result of the Oklahoma City bombing, federal building designs were changed in order to make the structures more blast-proof. As a result of the late design changes, Clark's labour costs were impacted. Clark used a measured mile analysis comparing the unaffected work prior to the design changes and the affected work after. Clark made adjustments to the measured mile due to the floor elevation and type of construction. The government characterized the Clark measured mile as a total cost method of pricing a claim. The board ruled that the measured mile analysis was the preferred method and a reasonable method to calculate lost labour and efficiencies. The board did selectively agree or disagree with some of Clark's measured mile adjustments.

5.4.1.2 Conclusion

The courts generally recognise the validity of loss of efficiency claims based on the measured mile analysis. While there are many methods of computing such damages, the use of a measured mile analysis, properly developed by a qualified expert, is the most reliable. The success or failure of a construction project often rides on the shoulders of labour productivity. Therefore it is incumbent upon the industry to educate and understand the basis for measurement and monitoring of labour productivity. If a project's planning and budget allow for the assignment of cost engineers and quantity surveys, the project has an extremely high probability of success. Also, should occurrences, problems, and/or events be introduced into a project, it is critical for management to recognise the importance and return on investment of employing onsite monitoring of the labour productivity. The measured mile technique provides the avenue for loss of productivity mitigation and labour impact analysis.

5.4.2 The 'Earned Value' approach

Current performance is the best indicator of future performance and therefore by using trend data it is possible to forecast programme and/or cost overruns at quite an early stage in a project. The most comprehensive trend analysis technique is the 'Earned Value' method.

Earned Value is an approach where you monitor the project programme, actual work and work-completed progress and value to see if a project is on track. Earned Value indicates how much of the time and budget should have been spent compared to the amount of work done to date.

Earned Value differs from the usual budget versus actual costs incurred model, in that it requires the progress and/or cost of work in progress to be quantified. This allows the project team to compare how much work has been completed against how much was expected to be completed at a given point.

The project manager needs to agree the project scope, create a Work Breakdown Structure [1] (WBS) and assign budget to each work package [2], the lowest level of the WBS. Next he/she will create a schedule showing the calendar time it will take to complete the work. This overall plan is baselined (this is the planned value) and used to measure performance throughout the project. As each work package is completed (earned) it is compared with planned value, showing the work achieved against plan. A variance to the plan is recorded as a time or schedule deviation.

It is necessary to obtain the actual costs incurred for the project from the organisation's accounting system. This cost is compared with the earned value to show an over- or underrun situation.

Earned Value provides the project manager with an objective way of measuring performance and predicting future outcomes. This can enable him/her to report progress with greater confidence and highlight any overrun earlier. This in turn enables the management team to make cost and time allocation decisions earlier than would otherwise be the case.

It is generally true that past performance is a good indicator of future performance, and as such Earned Value is a very useful tool for predicting the outcome of projects in terms of time to completion, cost to completion and expected final costs.

Earned Value is also known as Performance Measurement, Management by Objectives, Budgeted Cost of Work Performed and Cost Schedule Control Systems.

In addition to assessing any slippages or cost overruns in a project in general terms, Earned Value Analysis (EVA) can be used to help determine whether or not a project is providing value for money. EVA concentrates on three basic parameters:

1. How much work *should* have been done so far.
2. How much money has *actually* been spent to progress the project so far.
3. What is the *value* of work that has been accomplished so far.

By comparing these values, assessments can be made about how efficient a project is and where problems may lie.

WORKED EXAMPLE

The 'Earned Value' method of analysis is considered to be one of the most reliable of the productivity analysis techniques. Therefore, for a worked example I am using the fundamentals of an investigation and report I prepared to demonstrate a contractor's entitlement to additional costs for employer responsible events and circumstances which caused loss of productivity.

Project details for the worked example

The project was the upgrading of a power station, which had been commissioned in the 1970s. The contract value was £38 million, and the contract period was 54 weeks.

Unfortunately, delays occurred on the project and the contractor's direct labour costs were substantially higher than had been recovered in their interim valuations. It was the contractor's view that this was as a result of disruption and delays to their labour force. I was commissioned by the contractor to investigate and report on the delays being caused and the under-recovery of the direct labour costs.

After a 4-week investigation, which included several meetings with the contractor's supervisory personnel, I produced a comprehensive report on the causes, i.e. the events and circumstances which had caused the loss of productivity in their labour force. My report also included supported calculations/assessments of the individual causes of loss of efficiency.

My report, of some 220 pages, was subdivided into 9 number sections. However, the substantive sections were sections F and G, titled:

Section F: Disruption.

Section G: Loss of Efficiency; Quantitative Review.

I now give extracts from section F my report, which explains to the reader my approach and investigation.

'Section F: Disruption', was subdivided as follows:

1. Introduction
2. Disruption and efficiency; how is it measured?
3. Common causes of loss of efficiency
4. Inefficient working relative to TQs and VIs, but not recorded
5. Recovery of the cost of loss of efficiency
6. Direct labour manhours expended from 9 April to 23 November 2007
7. Methods of establishing lost efficiency
8. Establishing the loss of efficiency for this project: 'Earned Value' technique

Under the subheading, *1. Introduction,* I included the following paragraphs:

F2 Let me state at the onset of this section of my Report that my approach was not to start with the sum of actual direct hours expended, and then 'create' efficiency factors for the various causes of loss of efficiency to justify the additional hours expended; my approach was entirely the opposite.

F3 Basically, my approach was to start with a 'blank sheet of paper', and then with the guidance of established industry publications on productivity and efficiency, use the project's contemporaneous records to calculate or assess factors and unproductive hours for the various individual causes of loss of efficiency.

F4 This is a proactive approach and involved many hours of detailed research, investigation and analysis.

F5 It is generally accepted that, subject to a contractor's contemporaneous project records, the best approach is to use recognised industry studies on efficiency and productivity in conjunction with qualified expert opinion; as opposed to the blind application of such studies in a vacuum, e.g. without regard to the facts of the particular project. In other words, application of a labour productivity study, by a qualified expert in a reasonable manner, is a reasonable means to calculate the impact on labour efficiency.

Under tsubheading 3, *Common causes of loss of efficiency,* I used the listing of common causes of loss of efficiency, as described earlier in this chapter, as a template/checklist in my investigation.

Of the list of 24 causes of poor labour efficiency, some 12 were considered to be primary causes on this project. These were:
- (2) Acceleration (directed or constructive)
- (3) Adverse or unusually severe weather
- (5) Changes, ripple impact, cumulative impact of multiple changes and rework
- (8) Crowding of labour or stacking of trades
- (9) Defective engineering, engineering recycle and/or rework
- (11) Excessive overtime

- (12) Failure to coordinate trade contractors, subcontractors and/or vendors
- (17) Overmanning
- (18) Poor morale of craft labour
- (19) Out of sequence work
- (20) Programme compression impacts on productivity
- (24) Untimely approvals or responses.

The remaining 12 causes were considered to be secondary causes on this project, and are addressed for specific work areas/components as appropriate.

I also included the following explanatory paragraphs in subsection 3:

F27 Once the first efficiency loss has been detected, the contractor should reconfirm the baseline estimate to ascertain that the project estimate is basically correct. In doing so, the contractor can ensure that the efficiency loss detected is not simply the result of comparing onsite productivity and efficiency to a flawed baseline.

F28 Once this effort is complete the root cause of the efficiency loss needs to be determined. If the causation is found to be something for which the owner is liable, recommended practice is to follow the mandates of the contract with respect to providing written notice to the appropriate party as soon as possible.

F29 Subsequently, recommended practice is to gather all necessary supporting documentation and file the lost productivity, or efficiency, claim as prescribed in the contract. Some contracts allow claim filing within a specified period of time after the notice of claim was filed whereas other contracts provide for claim filing within so many days after the event or circumstance has passed. Regardless, the contractor seeking to recover lost productivity and/or efficiency costs should follow the mandates of the contract as closely as possible."

Sub-section 4, Inefficient working relative to TQs and VIs, but not recorded", included the following:

F30 From my investigation, it is clear, and confirmed by 'XXX' personnel, that the hours recorded on 'XXX's' 'labour daywork charge sheets' are the direct labour hours for carrying out the rework/rectification work only.

In essence, the hours recorded on the 'labour daywork charge sheets', and the subsequent VI, do not take into account the unproductive direct labour time spent on:

(1) finding the error/mistake;
(2) raising the problem with the 'XXX' supervisor;
(3) waiting whilst the 'XXX' supervisor queries this with 'ZZZ', either verbally or by TQ;
(4) if waiting for a 'ZZZ' response, then the time spent in transferring to another work operation and its learning curve;
(5) transferring back to the delayed work activity and starting the re-work (including its learning curve).

F31 There is no recorded direct labour time for the consequences of the occurrence of drawings or design information which is erroneous, ambiguous or unclear and resulted in 'XXX' raising an official Technical Query (TQ). When these situations occur and the direct labour is disrupted, probably to the same extent, either in part or total, as outlined in paragraph F27 above, then the unproductive direct labour time is not isolated and recorded on forms such as a 'labour daywork charge sheet'.

F32 For example, if a man works a 10-hour shift with no time spent on VI work, then 10-hours are recorded for that day on his weekly timesheet.

If, however, there is an event or circumstance, such as erroneous, ambiguous or unclear design information, then the following takes place regarding the recording of his time spent on his 10-hour shift:

(i) If time is spent on rework or rectification work, then this time is isolated and recorded separately on a 'labour daywork charge sheet', and these are subsequently forwarded to 'ZZZ' by way of a Variation Instruction (VI).

(ii) However, the time spent on 'finding' the problem, raising the technical query, etc. as described in parts 1 to 5 of paragraph F27 above, is not isolated and recorded separately, but included in the 10 hours recorded on the weekly timesheet.

Therefore, in this example, say the man spends 2 hours on the rework or rectification then this is recorded on a 'labour daywork charge sheet'. The remaining 8 hours, of his 10-hour shift, is recorded on his weekly timesheet for that day; there being no separate record of the disrupted and unproductive time spent on finding the problem and the subsequent consequences (see the 5 steps in paragraph F27).

F33 As the time involved in carrying out the unproductive work operations described in paragraph F27 is not recorded separately in the project's contemporaneous records, I have discussed the time involved for each of the '5 steps' with 'XXX' personnel, and the outcome of these discussions, together with my own experienced opinion, is as follows,

(1) "finding the error/mistake"; 20 minutes.
(2) "raising the problem with the 'XXX' supervisor"; 10 minutes.
(3) "waiting whilst the 'XXX' supervisor queries this with 'ZZZ', either verbally or by TQ"; 10 minutes.
(4) "If waiting for a 'ZZZ' response, then the time spent in transferring to another work operation and its learning curve"; 20 minutes.
(5) "transferring back to the delayed work activity and starting the re-work (including its learning curve)"; 10 minutes.

F34 This gives a total of 70 minutes disruption, or unproductive time for each 'event or circumstance'. Therefore, for example, if the 'event or circumstance' involved a gang of 4 men then there would be a total of 280 minutes, or 4 hours and 40 minutes unproductive time which at present is included in the general hours recorded on the weekly direct labour timesheets. For a 6-man gang of labour, the total would be 7 hours unproductive time for each 'event or circumstance'.

Subsection, Recovery of the cost of loss of efficiency, included the following:

F35 Based upon the definition of productivity set forth earlier in this section, and a review of the causation factors, loss of efficiency is

'The increased cost of performance caused by a change in the contractor's anticipated or planned resources, working conditions or method of work'.

F36 While the general cause(s) of a loss of efficiency may be easy to speculate upon (at least in hindsight), a contractor seeking to be compensated for a cost increase must first demonstrate entitlement; that is, a contractual right to recover damages. Second, the contractor must sufficiently prove causation, the nexus between entitlement and damages. The resulting damages, i.e. increased cost, is an outgrowth of the change in Output/Input. Loss of efficiency is the difference between baseline efficiency and that actually achieved.

F37 Baseline efficiency can be determined by measurements of input and output in unimpacted or the least impacted periods of time on the project. When this data is not available, estimated or analytically determined baseline efficiency may be substituted.

F38 I have researched and reviewed various published industry studies on productivity and efficiency, and in my view the studies most appropriate for the conditions and circumstances relating to this project are,

(i) 'Effects of Accelerated Working, Delays and Disruption on Labour Productivity', published by The Chartered Institute of Building (CIOB).

(ii) Various studies published by the Mechanical Contractors Association of America (MCA).

Various bulletins published by the Association for the Advancement of Cost Engineering (AACE).

F39 From the above studies, I now detail their recommendations relative to the 12 number 'primary' causes of poor labour productivity as described in paragraphs F19 and F20 above.

1. 'Acceleration' (cause ii); the CIOB is very informative on this cause, and their research shows an efficiency loss of 10% when working a 6 × 9 hour day work pattern, and a loss of 18% when working a 6 × 10 hour day work pattern.

   In my opinion, an efficiency loss of 12% is appropriate for this project.

2. 'Adverse or unusually severe weather' (cause iii); 'XXX' recorded separately the labour hours impacted by inclement weather. I have used these recorded inclement weather hours for my loss of efficiency assessment due to this cause.

3. 'Changes, ripple impact, cumulative impact of multiple changes and rework' (cause v); the MCA publication gives a range of efficiency loss from 10% to 20%, with an average of 15%.

   I consider that an efficiency loss of 15% is appropriate for this project.

4. 'Crowding of labour, or stacking of trades' (cause viii); the CIOB publication shows efficiency loss up to 45% as a result of this cause.

In my view, 45% is too high for the circumstances on this project, and I consider 15% to be more appropriate.

5. 'Defective engineering, engineering recycle and/or rework' (cause ix); none of the studies reviewed give any efficiency loss factors for this cause.

   Therefore, I consider a nominal 10% efficiency loss factor should be used for this cause.

6. 'Excessive overtime'(cause xi); the CIOB publication shows an efficiency loss of 12% when working a 6 × 9 hour day work pattern for a four week period for this cause. The MCA publication gives a range of efficiency loss from 10% to 20%, with an average of 15%.

   I consider the efficiency loss of 12% is appropriate for this project.

7. 'Failure to coordinate trade contractors, subcontractors and/or vendors' (cause xii); none of the studies reviewed give any efficiency loss factors for this cause.

   Therefore, I consider a nominal 10% efficiency loss factor should be used for this cause.

8. 'Overmanning' (cause xvii); the CIUOB publication quotes various studies that have been carried out and gives an efficiency loss of up to 40% as an average.

   Based on this data available, I consider an efficiency loss of 20% should be used for this cause on this project.

9. 'Poor morale of craft labour' (cause xviii); the MCA publication gives a range of efficiency loss from 5l% to 30%, with an average of 15% for this cause.

   However, I consider this average to be somewhat high, and am using an efficiency loss factor of 8% for this cause.

10. 'Out of sequence work' (cause xix); the CIOB study is based on return visits to the same workface, but based on a house project by a single tradesman. The result of this sampling is not considered appropriate to the circumstances on this project.

    As no other studies on this cause are available, I consider a nominal 5% efficiency loss factor should be used for this cause.

11. 'Programme compression impacts on productivity' (cause xx); again, none of the studies reviewed give any efficiency loss factors for this cause.

    Therefore, I consider a nominal 5% efficiency loss factor should be used for this cause.

12. 'Untimely approvals or responses' (cause xxiv); a somewhat contentious cause, and I have found no loss of efficiency factors in the studies reviewed.

    Therefore, I consider a nominal 5% efficiency loss factor should be used for this cause.

F40 The total of the efficiency loss factors for the above 12 primary causes is 112%. Obviously it is incorrect to apply this figure to the expended direct labour manhours on the delayed work areas/components as many of the causes will be overlapping.

F41 If one looks at the average loss factor for the 11 number causes, then a figure of 10% emerges. However, I consider this is a too low efficiency loss figure to apply in generality, and propose to uplift this figure by 100%, giving a global lost efficiency factor of 20%, which I will use in the appropriate circumstances.

Subsection 6, Direct labour manhours expended from 9 April to 23 November 2007, included the following:

"F42 In appendix F1 is a table showing the correlation between the Contract Programme 'Work Area/Component' numbers and 'XXX''s 'Work Package' numbers.

F43 From 'XXX''s contemporaneous records, the total direct labour manhours expended from 9 April 2007 to 23 November 2007 were 202,183.
This total is the combination of the following sub-totals,
Work Packages, i.e. base contract work   229,689
Variation Instructions   36,530
Inclement Weather   15,884
Unproductive work (relating to TQs,VIs, etc)   nil

F44 Appendix F2 shows the direct labour manhours for base contract work only spent on each Work Package, and subdivided into the three time periods, tranches 2, 3 and 4.

Subsection 8, Establishing the loss of efficiency for this project: 'Earned Value' technique", included the following:

"F50 At the end of Tranche 4, 'XXX' had completed some 9 number work areas/components. The following table shows the planned and actual direct labour manhours for each of these 9 number work areas/components.

		Manhours		
		Planned	Actual	Variance
1090	Duct Pre-Assembly: Outage U3	13,029	17,608	4,579
1180	Steelwork: Outage U3	6,750	2,603	−4,147
1190	Steelwork: Outage U4	3,972	987	−2,985
1200	Steelwork: Outage U2	5,956	2,332	−3,624
1210	Duct Installation: Outage U3	2,560	2,638	78
1220	Duct Installation: Outage U4	3,200	2,622	−578
1230	Duct Installation: Outage U2	2,009	751	−1,258
1410	Duct Pre-Assembly: Pre GGH U3	??	??	
1420	Duct Pre-Assembly: Non Tie-In U3	??	??	

F51 As can be seen from the above table, the actual direct labour manhours for 5 number work areas/components was less than planned, i.e. as noted on the April Contract Programme 'ACC1'.

F52 From my review of the contemporaneous project records I have noted that there were very few, if any, events which delayed 'XXX''s onsite direct labour working on these 5 number work areas/components.

F53 This, in my opinion, demonstrates that where there are few, if any, delaying events impacting work area/components, then the planned manhour budget was/is achievable.

F54 Compare this with the 2 work areas/component where the expended manhours were in excess of those planned, we see the following overruns and productivity output.

		Manhours		
		Planned	**Actual**	**Variance**
1090	Duct Pre-Assembly: Outage U3	13,029	17,608	−4,579
1210	Duct Installation: Outage U3	2,560	2,638	−78

F55 Putting aside the 'Duct Installation: Outage U3' work area, for which the variance in manhours is minimal, I comment on the other work area 'Duct Pre-Assembly: Outage U3' as follows.

F56 This work area/component was impacted by several delaying events and circumstances, such as 29 number TQs and 44 number VIs.

F57 In addition, this work area was impacted by many of the primary causes of productivity loss as described earlier in this section. From my review of the various industry studies on productivity loss, my analysis has resulted in a global productivity loss factor of 20%.

F58 When this global productivity factor is applied to the actual manhours for 'Duct Pre-Assembly: Outage U3', the planned manhours will be increased from 13,029 to 15,635. As noted earlier, the actual direct labour manhours spent on this work area/component was 17,608; an increase of 1,973 over the 'revised planned' manhours.

F59 Whilst this is some 12% more than the revised plan, I consider this not to be excessive or exceptional based on the circumstances encountered on this project and that a generalised global productivity loss factor has been used for the calculation.

F60 It is certainly not, in my opinion, an example or demonstration of inadequate supervision or management by 'XXX'.

F61 However, in my opinion, using the findings and review for only one work area/component, i.e. 'Duct Pre-Assembly: Outage U3', is an insufficient sample for me to base and to give opinion on for the other work areas/components.

F62 In the next section of my Report, section G, titled 'Loss of Efficiency; Quantitative Review', I give my findings from the 'Earned Value Analysis' technique I have carried out.

Section G: Loss of Efficiency; Quantitative Review", was subdivided as follows:

1. General
2. Tranche 2; from 9 April 2007 to 24 June 2007
3. Tranche 3; from 25 June 2007 to 21 September 2007
4. Tranche 4; from 22 September 2007 to 23 November 2007
5. Summary

Under subheading 1, *General*, I included the following paragraphs:

G1 Using the 'Earned Value Analysis' technique, as described in paragraph F45, I now present the results and findings in respect of the loss of efficiency of the direct labour on this project.

G2 The tables for each of the Tranches 2, 3 and 4 show the earned value of direct labour manhours based on actual progress achieved, which is obtained from 'ZZZ's progress updates, compared with the actual direct labour manhours expended.

Subsection 2, *Tranche 2 (9 April to 23 June 2007*, included the following paragraphs:

G3 The earned value and actual manhours expended on each of the work areas/components progressed in this tranche are as follows.

	Tranche 2	Manhours, in tranche		
		Earned Value	Actual	Variance
1000	Absorber Unit 3	5,551	14,927	9,376
1010	Absorber Unit 4	3,749	6,386	2,637
1020	Absorber Unit 2	2,493	4,139	1,646
1090	Duct Pre-Assembly: Outage U3	9,511	9,370	−141
1100	Duct Pre-Assembly: Outage U4	12,341	9,209	−3,132
1180	Steelwork: Outage U3	2,835	1,115	−1,720
1210	Duct Installation: Outage U3	358	1,541	1,183
1270	Steelwork: U3	840	262	−578
1280	Steelwork: U4	1,113	262	−851

G4 As can be seen from the table for Tranche 2, the actual manhours for 5 number of the 9 work areas/components is less than the earned value hours. Indeed, for 'Steelwork: Outage U3' the actual hours are less than half the earned value hours.

G5 Appendix xx contains the following documents which are applicable to the work areas/components in this tranche.
   (i) Schedule of Technical Queries (TQs)
   (ii) Schedule of Variation Instructions (VIs)
G6 A review of the contemporaneous project correspondence and records shows the following events and circumstances which caused the excessive manhours and loss of efficiency on the 4 number work areas/components whose actual hours exceeded their earned value. For each work area/component I give the results of my investigation, findings and opinion in respect of the individual causes of loss of efficiency.

Absorber Unit 3
G7 During this tranche the following events/circumstances occurred which would have caused disruption and loss of productivity to this work area/component.
   (i) 22 number Technical Queries (TQs). Whilst most of the TQs raised by 'XXX' were answered by 'ZZZ' within 3 days, there were 6 number where 'ZZZ' took longer than 3 days to provide 'XXX' with a reply. These were,
   - TQ 059, raised on 17 April but reply not received until 24 April 2007, for "Compensate pad for manhole to be cut".
   - TQ 069, raised on 24 April but reply not received until 28 April 2007, for "Possible clash between inter nozzle and vertical external stiffeners".
   - TQ 071, raised on 26 April but reply not received until 1 May 2007, for "Missing drawing for absorber 3 from nozzle up to head".
   - TQ 086, raised on 13 May but reply not received until 18 May 2007, for "Missing parts for stiffener ring EL34".
   - TQ 087, raised on 14 May but reply not received until 18 May 2007, for "Supports for spray manifold".
   - TQ 105, raised on 23 May but reply not received until 30 May 2007, for "Additional part on stiffener on stiffener ring _25,100 ?".
   (ii) 7 number Variation Instructions (VIs). None of these VIs appears to be as a consequence of any of the 22 number TQ's raised by 'XXX' for this work area/component. I have noted that all these VIs contained additional direct labour hours totalling 69 hours, but only 1 number had a work content of more than 20 manhours.

This was,

   - VI 659, dated 28 May 2007, for "As per dwg FF-0105-U3 WD-034-SVG rev2, parts 151 & 153 should have a 150 mm cut out. They have been manufacture without the cut out". This VI has 30 manhours recorded on 'XXX"s 'labour daywork charge sheet' number 80.

G8 'Adverse or unusually severe weather', cause 3) as detailed in paragraph F19.

During this tranche, the 'XXX' records show that for all work areas/components some 6,519 direct labour hours were booked for inclement weather. To calculate the 'inclement weather hours' for only this work area/component I am using the ratio of 'inclement weather hours' / 'work package hours plus VI hours'; which is (6,519 / 50,079 + 1,119) 0.13. Therefore, my assessment of the 'inclement weather' hours for this work area/component is (14,927 × 0.13) 1,940 hours.

This amount of labour hours is already isolated and recorded separately from the base contract hours recorded against the work packages for this work area/component. However, there would, in my opinion, have been a further loss of efficiency to these recorded hours by this cause. The loss of efficiency factor of 13% is for inclement weather, which is, in my opinion and experience, high for the months of April to June. However, in my view a loss of efficiency factor of 10% should be applied for this cause of 'adverse or unusually severe weather' to this recorded work package hours in this tranche which covers the time period of April to June 2007.

This 10% loss of efficiency factor equates to 1,493 hours.

G9 'Changes, ripple impact, cumulative impact of multiple changes and rework', cause 5) as detailed in paragraph F19.

In paragraph F29, I outline my reasoning for allowing a 15% loss of efficiency factor for this cause. However, it is my view that this loss of efficiency factor does not properly explain the disruption caused to this work area/component. I have, therefore, carried out a more detailed assessment as follows

As explained in paragraph F28 earlier,

"There is no recorded direct labour time for the consequences of the occurrence of drawings or design information which is erroneous, ambiguous or unclear and resulted in 'XXX' raising an official Technical Query (TQ). When these situations occur and the direct labour is disrupted, probably to the same extent, either in part or total, as outlined in paragraph F27 above, then the unproductive direct labour time is not isolated and recorded on forms such as a 'labour daywork charge sheet.'"

For my calculation of this unproductive time I am using the assessment of the time taken for steps 1 to 5 of the action taken when finding an engineering 'problem' on site. The details are given in paragraph F30; and the assessment of time for steps 1 to 5 is 70 minutes. Therefore, based on an average gang size of 6 men, for the 22 number TQs in this tranche, this gives a total of 154 hours.

G40 'Defective engineering, engineering re-cycle and/or rework', cause 9) as detailed in paragraph F19.

In paragraph F29, I outline my reasoning for allowing a 10% loss of efficiency factor for this cause. However, again it is my view that this loss of efficiency factor does not properly explain the disruption caused to this work area/component. I have, therefore, carried out a more detailed assessment as follows

Loss of Productivity

As explained in paragraph F27 earlier, the hours recorded on 'XXX''s 'labour daywork charge sheets' are those for carrying out the rework/rectification work only, and do not include direct labour hours, i.e. unproductive time, which would have been spent on identifying and resolving what was to be done, i.e. the rework or rectification work necessary.

For my calculation of this unproductive time I am using the assessment of the time taken for steps 1 to 3 of the action taken when finding an engineering 'problem' on site. The details are given in paragraph F30; and the assessment of time for steps 1 to 3 is 40 minutes. Therefore, based on an average gang size of 6 men, for the 7 number VIs in this tranche, this gives a total of 28 hours.

G41 'Failure to coordinate trade contractors, subcontractors and/or vendors', cause 12) as detailed in paragraph F19.

There are numerous instances of 'ZZZ' supplied materials not being delivered to site on time in this tranche. This fact, coupled with the delay and disruption caused through lack of scaffolding, by a 'ZZZ' employed subcontractor, has persuaded me that there was a loss of efficiency due to this cause. In paragraph F29, I outline my reasoning for allowing a 10% loss of efficiency factor for this cause.

In my opinion, this 10% loss of efficiency factor should be applied to this work area/component in this tranche. This factor equates to 1,493 hours.

G42 'Overmanning', cause 17) as detailed in paragraph F19.

'XXX''s planned labour force for this work are/component was an average of 22 men per day; whilst the actual number of men working on this area in this tranche was an average of 19 men per day. This is slightly less than the planned average, therefore, no loss of efficiency in respect of this particular cause.

G43 However, on this work area/component I consider there would be some impact on efficiency by the combined effect of the above causes together with the following causes,
  'Acceleration', (cause 2), e.g. second shift.
  'Crowding of labour or stacking of trades', (cause 8).
 'Excessive overtime', (cause 11).
  'Poor morale of craft labour', (cause 18).
  'Out of sequence work', (cause 19).
  '*Programme* compression impacts on productivity', (cause 20).
 'Untimely approvals or responses', cause (24), e.g. 'ZZZ' responses to TQs.

Therefore, in my opinion, the global loss of efficiency factor of 20%, as evaluated in paragraph F31 should also be applied to this work area/component. This factor equates to 2,985 hours.

G44 To summarise for the work area 'Absorber Unit 3', my reasoned opinion, calculations and assessments result in the following unproductive direct labour hours which are included in 'XXX''s records under the work packages for this work area.

(i)	'Adverse or unusually severe weather', (cause 3)	1,493 hours
(ii)	'Changes, ripple impact, cumulative impact of multiple changes and rework', (cause 5)	154 hours
(iii)	'Defective engineering, engineering re-cycle and/or rework', (cause 9)	28 hours
(iv)	'Failure to coordinate trade contractors, subcontractors and/or vendors', (cause 12)	1,493 hours
(v)	'Overmanning', (cause 17)	nil
(vi)	Other causes, as highlighted in paragraph G42	2,985 hours
	Total	*6,153 hours*

G45 As noted earlier, this 6,153 of unproductive hours has been included in 'XXX's records under the work packages for this work area. As the 'variance hours', i.e. the work package hours in excess of the earned value hours in this tranche was 9,376 this leaves some 3,223 direct labour manhours for which I can find no explanation.

In this sub-section, similar detailed analyses and findings were given for the other three work areas/components whose actual hours exceeded their earned value.

    Absorber Unit 4
    Absorber Unit 3
    Duct Installation: Outage U3.

Subsection 3, *Tranche 2 (24 June to 20 September 2007)*, contained similar scenarios and analyses for each of the 14 number work areas/components whose actual hours exceeded their earned value.

Subsection 4, *Tranche 3 (21 September to 23 November 2007)*, also contained similar scenarios and analyses for each of the 12 number work areas/components whose actual hours exceeded their earned value.

The final subsection, *5. Summary*, contained the following:

G296 A summary of the results of my investigations for each work area/component is given in appendix G4. A brief explanation of the columns for each of the three tranches is given below.
    (i) 'Variance'; this shows the direct labour hours derived from the actual hours recorded against the work packages compared with the earned value hours, which are based on the April Target Cost budget hours and the percentage progress achieved in the tranche.

(ii) 'Unprod.'; are the unproductive hours from my loss of efficiency review and assessment as detailed in the earlier part of this section.
(iii) 'Rev. Var.', i.e. 'variance' minus 'unprod' hours.

G297 I now review the results and give my views and opinion.

Absorbers

G298 This group contains 6 number work areas/components. For 2 number work areas/components, no progress has been made as at 23 November 2007. These were,
(i) Absorber Internals U4 (1040)
(ii) Absorber Internals U2 (1050).

G299 The total of the revised variances for the three Absorbers, (1000), (1010) & (1020) is 25,155 hours. This is a high amount, but these results do not take into account the following:
(a) There are 14 number VIs which are shown as 'no cost yet'; hence no hours are included in the assessment.
(b) 'XXX' still have to carry out a material quantity re-measure. A partial re-measure and assessment carried out by 'XXX' in April 2007 highlighted that there was an additional 79 tonnes in the Absorbers, which equates to almost 2,000 manhours.

G300 However, when the additional hours for these two items are finalised, the variance for the Absorbers will still be high, probably in the region of 22,000 manhours.

G301 In my view, this high variance is probably due to an under-estimation of the hours per tonne for erection of the Absorbers. The Absorbers in 'ZZZ's 'Scope of Erection' document was priced by 'XXX', back in November 2006, at 24.5 hours per tonne.

G303 With regard to Absorber Internals U3 (1040), the table in appendix G4 shows a variance of 981 hours for tranche 3. However, the variance for tranche 4 was (378) hours, i.e. the actual hours were less than the earned value (EV) hours. This leaves a variance of 603 hours, which equates to a productivity factor of 0.81 for this work area/component.

Steelwork

G304 This group contains 7 number work areas/components. For 1 number work areas/components, no progress has been made as at 23 October 2007. This was,
(i) Steelwork : U2 (1290).

G305 With regard to Steelwork: Outage U3 (1180), the table in appendix G4 shows a variance total of (4,160) hours for tranches 2 & 3, i.e. the actual hours were

> less than the earned value (EV) hours. Therefore, there is no productivity loss for this work area/component.
>
> G306 For Steelwork: Outage U4 (1190), the table in appendix G4 shows a variance of (2,985) hours for tranche 3, i.e. the actual hours were less than the earned value (EV) hours. Therefore, there is no productivity loss for this work area/component.
>
> G307 For Steelwork: Outage 02 (1200), the table in appendix G4 shows a variance of (3,609) hours for tranche 3, i.e. the actual hours were less than the earned value (EV) hours. Therefore, there is no productivity loss for this work area/component.
>
> G308 With regard to Steelwork: U3 (1270), the table in appendix G4 shows a variance total of 386 hours for tranches 3 & 4. However, the variance for tranche 2 was (514) hours, i.e. the actual hours were less than the earned value (EV) hours. As this is more than the combined variance of 386 hours for tranches 3 & 4, there is no productivity loss for this work area/component.
>
> G309 With regard to Steelwork: U4 (1280), the table in appendix G4 shows a variance of 135 hours for tranche 3. However, the variance total for tranches 2 & 4 was (1,270) hours, i.e. the actual hours were less than the earned value (EV) hours. As this is more than the variance of 135 hours for tranche 3, there is no productivity loss for this work area/component.
>
> G310 For Steelwork: Common (1200), the table in appendix G4 shows a variance total of (1,287) hours for tranches 3 & 4, i.e. the actual hours were less than the earned value (EV) hours. Therefore, there is no productivity loss for this work area/component.
>
> G311 To summarise the 'Steelwork' group; out of 7 number work areas/components, there is no productivity loss for any work area/component.

Similar views and opinions were given for the other work areas/components.

As can be seen, a very detailed investigation and analysis has to be carried out to demonstrate loss of productivity. To put forward a claim for reimbursement of costs from the employer/client the analysis has to demonstrate causation by linking the losses to factual events and circumstances for which the employer/client is contractually responsible.

Chapter 6
Acceleration and Mitigation

6.1 Acceleration

'Acceleration' tends to be bandied about as if it was a term with a precise meaning, but this is not the case.

The Society of Construction Law's Delay & Disruption Protocol defines 'acceleration' in Core Statement 20 of the Protocol, as:

> "Where the contract provides for acceleration, payment for the acceleration should be based on the terms of the contract. Where the contract does not provide for acceleration but the Contractor and the Employer agree that accelerative measures should be undertaken, the basis of payment should be agreed before the acceleration is commenced. It is not recommended that a claim for so-called constructive acceleration be made. Instead, prior to any acceleration measures, steps should be taken by either party to have the dispute or difference about entitlement to EOT resolved in accordance with the dispute resolution procedures applicable to the contract (see Guidance Section 1.18)."

6.1.1 What is acceleration?

On a construction or engineering project, acceleration is where work is carried out more quickly than previously planned. It usually occurs in one of two forms. Firstly, it occurs when a contractor or subcontractor is required to carry out increased, additional or delayed work within the same time period without the benefit of being given an extension of time.

Secondly, acceleration is closely related to disruption. It is only in recent years that standard forms of contract have made allowance for the employer to issue instructions to accelerate the works.

The reasons for acceleration usually fall into one of the following categories:

1. **By agreement or instruction**. By agreement between the parties or, if the contract so provides, on the instruction of the architect.

Practical Guide to Disruption and Productivity Loss on Construction and Engineering Projects, First Edition. Roger Gibson.
© 2015 John Wiley & Sons, Ltd. Published 2015 by John Wiley & Sons, Ltd.

2. **Unilateral acceleration**. Unilaterally on the initiative of the contractor, often categorised as 'mitigation' by the contractor or as 'using best endeavours' by the employer.
3. **Constructive acceleration**. Constructive acceleration is where the contractor argues that he has no real alternative in the circumstances.

Acceleration occurs when a contractor is required to perform its work in less time than originally planned. The party liable for the cost of acceleration is the party responsible for the underlying delay and/or the party deciding to accelerate.

For example: the contractor is required to install 5,000 m of pipework in 30 days. If the employer later requires the contractor to install 5,000 m of pipework in 20 days, or install 7,000 m of pipework in 30 days, then the contractor was 'accelerated'. This is an example of '*instructed acceleration*'.

'*Constructive acceleration*' occurs when a contractor encounters an excusable delay during the carrying out of the contract work, such as design changes, late information or employer-caused delays. Under the contract, the contractor is entitled to an extension of time. If the contractor is not granted a time extension then he is constructively accelerated in its obligation to meet the contract completion date.

The Society of Construction Law's 'Delay and Disruption Protocol' defines acceleration, under its Core Principle clauses, as follows:

> 20. Acceleration.
> Where the contract provides for acceleration, payment for the acceleration should be based on the terms of the contract. Where the contract does not provide for acceleration but the Contractor and the Employer agree that acceleration measures should be undertaken, the basis of payment should be agreed before the acceleration is commenced. It is not recommended that a claim for so-called constructive acceleration be made. Instead, prior to any acceleration measures, steps should be taken by either party to have the dispute or difference about entitlement to EOT resolved in accordance with the dispute resolution procedures applicable to the contract.

6.1.2 Acceleration under the contract

6.1.2.1 JCT Standard Building Contract 2005

No mention is made of 'acceleration' in the main body of the contract, except at clause 1.1, where the term 'Acceleration Quotation', is defined as meaning:

> ".... a quotation by the Contractor for an acceleration in the carrying out of the Works or a Section made under paragraph 2 of Schedule 2."

The primary purpose of Schedule 2 is to set out the procedure for 'Variation Quotations'. The architect, or contract administrator, may instruct the contractor to provide variation quotations by virtue of clause 5.3.1 of the main terms of the contract:

"If the Employer wishes to investigate the possibility of achieving practical completion before the Completion Date ... the Architect/Contract Administrator shall invite proposals from the Contractor in that regard."

The 'Acceleration Quotation' must identify the amount of time that can be saved and the amount of the adjustment to the 'Contract Sum' that the contractor would require. The quotation must include direct costs, consequential loss and expense and an allowance for the cost of preparing the quotation.

6.1.2.2 NEC3 Form of Contract

Acceleration is referred to in Core Clause 36 of the June 2005 edition of the NEC3 standard form of contract.

Under clause 36.1 the project manager may instruct the contractor to submit an acceleration quotation. As with the JCT form, the stated aim of acceleration is to achieve completion before 'the Completion Date". The 'Completion Date' may be the original date stated in the Contract Data (the final section of the NEC form) or a later revised date arising out of an extension of time award.

Unlike the JCT acceleration clause, it is not for the contractor to state what acceleration it can achieve; under clause 36.1 of the NEC it is the project manager who informs the contractor of the revised date, or dates, that it is required to achieve.

Following receipt of an instruction the contractor must provide a quotation and a revised programme showing how it can achieve the early completion date(s). The contractor may decline to quote but, if it does, it must state why (clause 36.2). Presumably the usual reason for declining to quote will be that the contractor considers the revised dates are not achievable.

6.1.2.3 ICE Conditions Contract, 7th Edition

Acceleration is referred to in clause 47(3) of the ICE Conditions Contract, 7th edition. Clause 47(3) provides that the employer may request the contractor to complete the works earlier than 'the time or extended time for completion prescribed by Clauses 43 and 44 as appropriate'. Clause 43 refers to the completion date in the Appendix to the Form of Tender (which is a standard document provided at the back of the contract); clause 44 refers to extensions of time.

If the employer requests the contractor to complete early and the contractor agrees, then 'any special terms and conditions of payment shall be agreed ... before any such action is taken'.

6.1.3 Acceleration by agreement or instruction

There should be no difficulty in obtaining payment where the contract administrator, in exercise of his powers under a contract, orders acceleration of the work or the employer and the contractor agree acceleration and a claim under the direct loss and expense clause is unnecessary.

Once the contract administrator instructs acceleration, it is clear that the contractor must be paid for it by the employer. It follows that where directed acceleration has been instructed, the contractor is entitled to be paid:

(a) The agreed rate for acceleration, if any rate has been agreed; or
(b) In the absence of an agreed rate, a reasonable rate for the acceleration measures, i.e. the contractor's actual costs plus a reasonable level of profit and overheads.

6.1.4 Unilateral acceleration

This is the situation where a contractor accelerates without any agreement with the employer or instruction from the architect. No pressure has been placed on him by the refusal of an extension of time; indeed, in this situation it may be that the contractor is reasonably confident of getting an extension of time. The reason for doing so may be in order to find work for operatives from another site which is drawing to a close. The result may be that some time is recovered and an extension of time is not required.

In most such cases, the contractor will find it difficult to contend that he was going other than 'using his best endeavours' to reduce delay. It is by no means clear, however, under what contract provision the contractor could be paid even if the architect agrees.

6.1.5 Constructive acceleration

This is an argument advanced by a contractor and is based on the architect's failure to give an extension of time to which the contractor believes he is entitled. A contractor will put more resources into a project than originally envisaged and then attempt to recover the value on the basis that he was obliged to do so in order to complete on time, because the architect failed to make an extension of time of the contract period. The problem faced by the contractor is that, in the absence of an extension of time, he may be faced with liquidated damages being levied against him. He has a stark choice: he can continue to work as planned and efficiently in the hope that he can later successfully demonstrate that he is entitled to an extension of time and that this will be granted.

Alternatively, he can accept, temporarily at least, that he is in default and take steps to mitigate the consequences of this temporary default by putting more resources on the project, and/or reorganising the works so as to finish by the date for completion.

An important question to be asked before this kind of argument can be entertained is the extent to which pressure is put on the contractor; the contractor's problem is one of causation. Where the architect fails to make an extension of time, either at all or of sufficient length, the contractor's route under the contract is adjudication or arbitration. If, as a matter of fact and law, the contractor is entitled to an extension of time, it may be said that he can confidently continue the work, without increasing resources, secure in the knowledge that he will be able to recover his prolongation loss and/or expense and any liquidated damages wrongfully deducted, at adjudication or arbitration. If he increases his resources, that is not a direct result of the architect's breach but of the contractor's decision.

In practice, it must be acknowledged that a contractor in this position may not be entirely confident; the facts may be complex and the liquidated damages high. Faith in the wisdom of an adjudicator or arbitrator may not be total. It may be cheaper, even without recovering acceleration costs, for the contractor to accelerate rather than face liquidated damages with no guarantee that an extension of time will ultimately be made. As a matter of plain commercial realism, the contractor may have no other sensible choice than to accelerate and take a chance as to recovery. Unless the contractor can show that the architect has given him no real expectation that the contract period will ever be extended and in those circumstances the amount of liquidated damages would effectively bring about insolvency, this kind of claim has little chance of success.

However, under the Housing Grants, Construction and Regeneration Act, a contractor now has the option to address the uncertainty at an early stage and not wait until after completion of the project. He can refer the architect's/contract administrator's refusal of his extension of time claim to an adjudicator during the course of the contract rather than to arbitration or litigation after completion of the project.

In the United States, a 'constructive acceleration' doctrine has been established to permit a contractor to claim his acceleration costs. The US doctrine, modified for the British construction scene, comprises a six-stage test of the following questions:

1. Is there a delay, resulting from a relevant event, that would entitle the contractor to an extension of time?
2. Has the architect/contract administrator been given notice of the delay in accordance with the contract?
3. Has the architect/contract administrator refused or failed to grant an extension of time?

4. Has the architect/contract administrator, or employer, acted in some manner that can be construed as an instruction to complete by the original or revised date for completion?
5. Has the contractor accelerated its performance?
6. Has the contractor incurred additional costs as a result?

Hudson's Building and Engineering Contracts refers to the concept of 'Constructive Acceleration' as follows:

> "In the United States, a highly ingenious type of contractor's claim, based on a 'constructive acceleration order' theory, has been accepted in the Court of Claims for government contracts in the not uncommon situation where an [architect/engineer], in a bona fide belief that the contractor is not entitled to an extension of time and is in default, presses a contractor to complete by the original completion date, and it is subsequently held that the contractor had been entitled to an extension of time. This is, however, a development of what, in any event, is a largely jurisdictional and fictitious doctrine of 'constructive change orders' (CCOs) developed by the Boards of Contract Appeals, and is not founded on any consensual or quasi-contractual basis which would be acceptable in English or Commonwealth Courts, it is submitted."

Typically, when a delay occurs in a project, the contractor often expedites progress through 'activity crashing' with respect to available float and time-cost relationships. In effect, the contractor prescribes overtime work and/or injects additional resources in order to shorten (crash) the duration of certain activities. While injecting additional resources can significantly increase project costs, prolonged overtime working may cause declines in productivity and performance, which may also generate rework.

6.1.6 Recovery of acceleration costs

Usually, if it can be shown that the acceleration has been caused by an event for which there should have been compensation, then there is no reason why the costs should not be recoverable as loss and expense and valued in the usual way under the contract. However, if they cannot be so valued, then it is possible that the claim can proceed on a quantum merit basis of the reasonable costs of the accelerated works.

In the following three sections we highlight and expand upon what we consider are four important aspects under the global heading of *Acceleration*.

Accelerated working will usually arise as a consequence of one of the following:
The employer instructs the contractor to complete the works quicker than planned.

The employer calls for additional work to be completed within the original period of the contract.

The contractor is in delay as a consequence of receiving information late or too slow.

The contractor is in delay as a consequence of his own mismanagement.

In each of the above cases, the same and sometimes more work has to be compressed into a shorter period of time.

Irrespective of which party is responsible, there will be a need to establish and quantify the effects. If the contractor's mismanagement is the cause he will need to balance the costs of accelerated working against the potential savings in liquidated damages. If the employer is at fault, the parties will need to agree the consequences so that the contractor can be equitably paid.

Both parties must also recognise that if the cause of delays is a slow flow of information, work cannot be accelerated unless the flow is speeded up. The problem is one of project management.

Although in some cases there is an express provision in the contract for the issue of an instruction from an employer to the contractor to accelerate, there have been very few cases in English courts on acceleration. However, acceleration in the American courts is quite common.

The American courts recognise two types of acceleration: directed and constructive. Bramble & Callahan, in *Construction Delay Claims*, use the US Army Corps of Engineers' definition of directed acceleration:

> "... [the] buying back of a time extension due the contractor under the terms of the contract in an effort to complete the work within the existing contract completion date."

In *Ascon Contracting Ltd v. Alfred McAlpine Construction Isle of Man Ltd*, Judge Hicks QC said: " ... acceleration not required to meet a contractor's existing obligations is likely to be the result of an instruction from the employer for which the latter must pay ...".

Constructive acceleration has been defined by the US Army Corps of Engineers as:

> "An act or failure to act by the [employer] which does not recognize that the contractor has encountered excusable delay for which he is entitled to a time extension and which required the contractor to accelerate his schedule in order to complete the contract requirements by the existing contract completion date."

A mere voluntary decision by the contractor to accelerate is not enough to justify a claim for constructive acceleration. There must be some element of "undue coercion by the employer", which compels the contractor to accelerate.

The necessary elements for constructive acceleration are:

1. There must be an excusable delay.
2. There must have been timely notice of the delay and a proper request for an extension of time.
3. The time extension request must either be postponed or refused.
4. The employer or other party must act *by coercion, direction, or in some other manner that reasonably can be construed as an orderto complete within the unextended performance period.*
5. The contractor must actually accelerate its performance and thereby incur added costs.

The refusal to grant a time extension after an excusable delay has been experienced and acceptable notice has been given is alone not sufficient to constitute an acceleration order. After a request for a time extension has been denied, something more is required before constructive acceleration is considered to have been ordered. The courts refer to the 'something more' as an express or implied demand to accelerate. The demand to accelerate has been found when a demand or order that the contractor complete the project on time has been made in connection with a refusal to grant a reasonable time extension. The acceleration order need not be phrased 'in explicitly mandatory terms'. For example, pressure on the contractor to employ larger crews and more equipment and communications with a contractor's bonding company in an effort to force the contractor to accelerate has been recognised as valid orders to accelerate. In addition, an expression of concern about lagging progress may have the same effect as an order to accelerate. Other common employer actions that may constitute an order to accelerate include:

- repeated assertions in project meetings (and project meeting minutes) that no extension of time will be permitted despite the occurrence of excusable delays;
- a statement of the urgency of completion on the original contract completion date, coupled with the employer's threatening to issue an unsatisfactory performance report regardless of the circumstances;
- insistence on meeting the original contract completion date despite late information delays that result in suspension of the work; and
- ignoring early objections by the contractor that employer-imposed durations are too short then later demanding completion on the employer-imposed date.

A case concerning constructive acceleration came before the TC Court in 2002, which was *Motherwell Bridge Construction Ltd v. Micafil Vakuumtechnik*. The case concerned the construction of a large steel vessel. Inter alia, the contractor said that the employer required a higher standard of welding than that stipulated in the contract. The change in welding was said to have resulted in slower progress on site. Despite this difficulty, the employer still demanded that the contractor achieve the original completion date. HHJ Toulmin accepted that the welding standard had become more onerous and allowed the contractor to recover certain acceleration costs as damages. In his judgment, HHJ Toulmin stateed:

> "[The contractors] were entitled to acceleration costs incurred as a result of trying to finish on time when delay was caused by [the employer]. They were also entitled to the cost of premium time incurred for the same reason."

Some of Judge Toulmin's main findings in regard to acceleration are as follows:

> "[The contractors] say that since they incurred these costs in attempting to comply with [the employer's] wish for the contract to be kept to time and against the background of [the employer's] refusal to grant appropriate extensions of time, they are entitled to be paid for the work which they did in trying to accelerate the work to keep up with the schedule.
>
> Mr Nicholl [project manager] wrote to [the employer] complaining about problems which delayed the start of work on the erection of the timetable on site at Erith. He noted a critical path delay of three weeks for which he asked for an extension of time. He said that by working additional hours two of the three weeks of delay could be made up.
>
> … [Another] letter went on at para 1.20 to set out the actual and projected costs as a result of their starting a night shift … and in incurring the cost of premium day shift working. This working was not within the costs for which [the contractors] could reasonably have been expected to estimate when they tendered for the contract.
>
> I am satisfied that they were incurred by [the contractors] in an attempt to recover time lost in completing the work in circumstances where [the contractors] were subject to significant penalties for delay if they failed to complete the work on time.
>
> [The contractors] … entirely reasonably required their employees to work in excess of the hours allowed for in their tender … in order to try to keep to the time schedule which had been imposed upon them and in respect of which they had not been given any relief by [the employer] to reflect the increase in work which they required to do or the difficulties of working on site. In those circumstances it seems to me reasonable that they should recover the sums which were paid to their employees … for work which they carried out at premium rates in order to try to keep to the required timetable.
>
> I find that this claim is allowable. If [the contractor's] staff were required to work excessive overtime in order to try to keep to a timetable which was always tight but was required in addition to accommodate the changes in design, [the contractors] are entitled to be recompensed for the consequences. I am satisfied that one of the direct consequences of excessive overtime was loss of productivity."

6.1.7 Costs recoverable under acceleration

Accelerating a project costs money. In addition to the direct cost of the work, acceleration also may result in a loss of productivity. It is recognised in the construction and engineering industries that acceleration efforts, such as working overtime and shift work, performing out-of-sequence work, stacking trades, and overcrowding on the project site contribute to reduced labour productivity. In addition, when a contractor adds labour, there may be a loss of labour productivity as new workers may not be familiar with the work or may require training before achieving normal levels of productivity.

It is interesting to note that a contractor typically does not have to prove that the acceleration effort was successful. It is normally necessary to show that it reasonably attempted to accelerate the work and that the acceleration efforts resulted in additional costs.

A lack of understanding often exists between contractors, employers and their consultants as to what may and what may not be included in acceleration claims.

There is no hard and fast formula for calculating acceleration costs. It is advisable that each situation should be individually assessed to determine what costs were sustained in the attempt to buy back time. Specific methods or a combination of methods which can be used to calculate acceleration costs exist. These are:

- The global or total cost approach
- The modified total cost approach
- The time impact methodology
- The measured mile approach, and
- Formula approaches.

Acceleration claims in construction and engineering projects very often result from the lack of a good control system. If there is no control system that can effectively register every change that occurs during the project execution disputes are likely to emerge. A typical checklist for an acceleration claim is as follows:

- A summary of items and amounts to be claimed
- Documents that support the claim
- A detailed analysis of how the amounts were calculated, and
- Legal and contractual support. Each standard form of contract has its own detailed requirements regarding record keeping, document control, notices, etc. The prime source of information for any claim between the parties is the contract and the specific requirements contained therein. The process of keeping project records should start during the tender process. Tender documents

are often used in disputes to help substantiate the costs that a contractor expects to incur on the project. Other project records such as project cost reports, daily logs and progress reports, daily payroll records, site instructions and related support documentation, minutes of meetings, project correspondence, documentation of design changes, photographs, etc. will be vital in substantiating an acceleration claim.

After a contractor has demonstrated that he was ordered to accelerate, he must then demonstrate that he reasonably attempted to accelerate. However, the contractor does not have to prove that he completed by the accelerated date in order for his acceleration claim to stand; he only needs to show that he incurred additional cost in a reasonable effort to accelerate.[Heading C below no bullet]

6.1.7.1 Changes to the programme

More efficient programming of the works may enable completion to be achieved earlier than would otherwise be the case, and this may be achieved without making any changes to the resource levels and/or the working hours. Progress can often be accelerated simply by altering the sequence of activities or by increasing the amount of overlap between activities.

However, on many new-build projects it will not be possible to alter the planned sequence but, for example, it may be possible to apply the first coat of emulsion paint to walls in advance of the planned period for decorations which should enable overall progress to be accelerated. Furthermore, on refurbishment works there are likely to be more options available for resequencing the programme, as there are generally more work faces available at any one time. Introducing or increasing the amount of overlap between activities may also boost progress. For example, it may be that a contractor's programme for the project shows the final activity of floor coverings commencing in week 42, immediately following the completion of decoration works. If the floor coverings take three weeks to complete, then it will be week 45 before the project will be completed. However, it may be that decorations in the first part of the building can be completed and sufficiently dried for floor coverings to commence in that area in week 40. The introduction of this two- week overlap should, all other things being equal, accelerate completion by two weeks.

6.1.7.2 Increased resources

The speed of progress of works on a construction site is often proportionate to the amount of labour, plant and/or supervision resources deployed. In general, the rate of progress increases if the amount of resources is increased and vice versa.

Therefore, one of the ways in which a contractor might accelerate a project is to increase the level of resources. For example, the contractor might increase the number of bricklaying gangs, mixers, forklifts (or other plant used to distribute materials around the site) and supervisors. It is likely such increases will boost output and speed up progress.

6.1.7.3 Working longer hours

The introduction of longer working hours should increase the overall rate of progress of works and enable completion to be achieved earlier than would otherwise have been the case.

To give an example; if site working hours were increased by one hour per day, or holiday periods worked, then it is likely that the rate of progress increase. Alternatively, progress on site may be improved simply by increasing the hours worked by operatives engaged in loading out materials. This can apply particularly on a large project with only one crane, where different trades often vie for 'hook time' to enable sufficient materials to be in place for operatives to progress the works.

A more dramatic change to working hours would be to introduce shifts. The introduction of one or two additional shifts in each 24-hour period ought to bring about very significant increases in output.

6.1.7.4 Introduction of additional temporary works measures

The rate of progress on a project may be increased simply by introducing additional temporary works measures. For example, progress may be expedited by the introduction of weather protection in the form of temporary roof sheets or temporary screens at openings. Such action would allow finishing trades, which require relatively dry conditions, to commence work earlier than would otherwise be the case and this should accelerate progress.

Alterations to scaffolding would also fall into this category. If more scaffolding is erected, or more boards provided to existing scaffold structures, then more work output ought to be achieved, providing this measure is combined with an increase in resources.

6.2 Mitigation

6.2.1 Introduction

In many jurisdictions throughout the world, within the area of civil law, a duty to mitigate damages is required. In essence, a duty to mitigate damages means that a potential plaintiff may not just sit back and let damages accrue without doing something to limit or minimise the accrual of damages. In many cases, a court will not allow a plaintiff to collect damages past the point at which he or she could have reasonably stopped the accrual.

Most forms of contract contain reference to a contractor's obligation to mitigate delays; including the effects of employer responsible delays.

JCT2005 contains the following at clause 2.28.6.1:

"... the Contractor shall constantly use his best endeavours to prevent delay in the progress of the Works or any Section, however caused, and to prevent the completion of the Works or Section being delayed or further delayed beyond the relevant Completion Date."

However, in many instances this leads to conflicting interpretations, for example, to what extent of 'best endeavours', or mitigation, is required. Beyond a certain level, e.g. if substantially more resources are mobilised, it could be interpreted as 'acceleration' to catch up for delays by others, rather than mere 'mitigation' which could lead to further claims for compensation.

The Guidance Notes of the Society of Construction Law's Delay and Disruption Protocol refer to Mitigation as follows:

1.5   Mitigation of delay
1.5.1   The Contractor has a general duty to mitigate the effect on its works of Employer Risk Events. Subject to express contract wording or agreement to the contrary, the duty to mitigate does not extend to requiring the Contractor to add extra resources or to work outside its planned working hours.
1.5.2   Note that the requirement in the UK Joint Contracts Tribunal (JCT) contracts for the Contractor to use 'best endeavours' to prevent delay in the progress of the works and prevent completion of the works being delayed beyond the completion date may place a higher burden on the Contractor than the normal duty to mitigate. In the event of Employer Delay, it is of course open to the Employer to agree to pay the Contractor for additional mitigation measures. See Guidance Section 1.13 below for mitigation of loss.
1.13   Mitigation of loss
1.13.1 The Contractor should do all it reasonably can to avoid the financial consequences of Employer Delay.
Guidance
1.13.2 The Contractor's duty to mitigate its loss has two aspects: first, the Contractor must take reasonable steps to minimise its loss; and secondly, the Contractor must not take unreasonable steps that increase its loss.
1.13.3 Most construction contracts include a requirement to the effect that the Contractor must do all it can to avoid, overcome or reduce delay.

Some forms actually make compliance with such provisions a condition precedent to the recovery of compensation or relief from LDs.

1.13.4 The limitations on the Contractor's obligations to mitigate Employer Delay are set out in Guidance Section 1.5. The Contractor does not have a duty to carry out any change in scope any more efficiently than the original scope. Neither is the Contractor obliged to expend money in order to mitigate the effect of an Employer Risk Event. If the Employer wishes the Contractor to take measures to mitigate the Employer Delay (whether by adding extra resources, by working outside its planned working hours or otherwise), the Employer should agree to pay the Contractor for the costs of those mitigation efforts.

1.13.5 It is the obligation of the Contractor to proceed with the works so as to complete on or before the completion date. However, the method, speed and timing of the activities forming the contract scope are generally left to the Contractor's discretion, subject to any stipulated prior process of acceptance of method and/or programme.

1.13.6 In the event that changes are made to the scope of the works, the Contractor has a similar obligation as to efficiency in relation to the changed scope as it has to the original scope.

The principles of mitigation are straightforward:

- The contractor cannot recover damages as may result from the employer's actions if it would have been possible to avoid such damage by taking reasonable measures.
- The contractor cannot recover damages which could have been avoided by taking measures greater than what might be considered reasonable.
- The contractor can recover the cost of taking reasonable measures to avoid or mitigate its potential damages.

The contractor is not obliged do everything possible. If that was the case, a successful claim for damages or loss and/or expense might be rare. The contractor need not do anything more than an ordinary prudent person in the course of his business would do.

Obviously, the contractor must have the right, in principle, to the damages before mitigation is relevant. It is important to understand that a failure to mitigate will not give rise to a legal liability, but it will reduce the damages recoverable to what they would have been had mitigating measures been taken. The extension of time clauses of many standard form contracts require the contractor to use its best endeavours to prevent delay occurring and to reduce the effects of a delay.

Where the contractor is claiming loss and/or expense under a standard form contract, it is for the employer to demonstrate that the claimant has failed to mitigate. Under most standard forms that is something which the architect would handle. In doing so, the architect would be entitled to request reasonable substantiating information.

A list of measures that can be used to mitigate these factors are:

1. **Preventive measures**. These are precautionary measures that are put in place as a defense to the inhibiting factors. Most of these measures are active measures that would be put in place during the planning stage of a project. For example, a preventive measure against the problem of design changes during cost and time of projects is to ensure that the project is designed to a great detail at the outset whilst a preventive measure for risk and uncertainty is to properly identify the project risks before the project starts and devise a strategy for managing them should they come to fruition.
2. **Predictive measures**. These may seem similar to preventive measures but they are not the same. Predictive measures are put in place in order to spot potential problems to the control process in the future so that they can be prevented from happening or be prepared for should they happen. Most of these measures actually utilise some tools or techniques to look into the current situation in a bid to spot potential future problems. For example, using 4D modelling (3D plus time dimension) to test how the plan (programme) will work out what is a predictive measure that could be used for the mitigation of complexity of works.
3. **Corrective measures**. These are measures that are utilised to mitigate the effect of the project control inhibiting factors by acting as a remedy. These measures are reactive measures that only act after the event. They may not be as effective as preventive or predictive measures but they aim to bring the situation back on track or at least 'stop the rot'. These measures have also been further classified as (a) corrective-preventive measures which are meant to correct and in the process prevent future problems, and (b) corrective-predictive measures which remedy the current situation but then go on to predict what the situation is going to be in the future using current information.
4. **Organisational measures**. These measures generally encompass practices that go wider than the actual control process but have an effect on project control. They are normally in place because of the company's belief, orientation, management style or philosophy. They have a tendency of not being specific to one project but would normally affect all projects being undertaken by the company as they reflect how the wider organisation works. A good example is the philosophy of the company in relation to partnering and collaborative working.

Some measures are fluid and can sometimes look as though they can be classified into more than one category depending on their actual usage during the project. Consequently this classification is not set in stone and should be seen as a first attempt at categorising the various good practices that can be used for mitigation of these leading project (cost and time) control inhibiting factors.

Appendix 1
Definitions and Glossary

This Appendix provides definitions and explanations for words and expressions commonly used in delay and disruption situations.

Acceleration: taking or planning active measures to complete work ahead of the project programme and/or to recover delays. Such action usually increases the overall cost of the project. See also 'Mitigation'.

Accepted Programme: a programme submitted by a contractor, or subcontractor, for the whole of the works for acceptance by the CA (or similar). Once accepted by the CA, it is known as the Accepted Programme.

Activity: an operation or process in a project that consumes time and also usually consumes or uses other resources, e.g. people, materials, equipment. An activity is a measurable element of the total project programme but, depending on the hierarchy or level of detail of the programme, may be divisible into smaller or more detailed activities. See also 'Task'.

Activity Duration: the time calculated or estimated to carry out an activity, generally taking into account a specific level of resource.

Activity ID: a unique code, usually alpha-numeric, that identifies each activity in a project.

Activity-on-Arrow Network: a network technique that uses arrows to represent activities. Preceding and succeeding activities join at nodes or events.

Activity-on-Node Network: a network technique that uses cells (generally symbolised by 'boxes') to represent activities. Also known as a Precedence Diagram.

Actual Dates: the dates relating to an activity when it started and/or finished.

Actual Duration: the duration that an activity took to complete.

Actual Progress: the amount of work that has been completed at a given point in time.

As-Built Dates: the actual start and finish dates for an activity.

As-Built Network: a network such that the activity durations, sequence and start and finish dates reflect the actual durations, actual start and actual finish dates. Dependencies and other constraints in the As-Built Network should be

---

Practical Guide to Disruption and Productivity Loss on Construction and Engineering Projects, First Edition. Roger Gibson.
© 2015 John Wiley & Sons, Ltd. Published 2015 by John Wiley & Sons, Ltd.

carefully considered to represent the actual dependencies and constraints encountered in the project and as such they result in the actual durations, actual start and actual finish dates of the activities.

As-Built Programme: a programme that represents the history of the project showing the actual start, actual finish and actual duration of the activities. The As-Built Programme does not necessarily have any logic links. It is usually in a barchart format.

As-Late-As-Possible: Timing or positioning of an activity in a programme at its Latest Start/Latest Finish dates such that there is no Free Float on the activity and the timing of other activities in the programme and overall duration of the programme is not affected.

Backward Pass: the procedure whereby the Latest Dates of activities in a network are calculated.

Barchart: a graphical chart on which activities are represented as bars drawn to a common timescale. Typically, a date scale is drawn across the top of the page and a list of activities down the left hand side of the page. Activity timing and durations are represented by horizontal bars. Additional information, such as resources, costs and dependencies are also often shown on the chart.

Baseline Programme: a fixed or record programme against which current or future activity is referenced. Often taken to mean the first or original plan, but can be reset (for instance following a change to the project scope) at which point the Reset Programme becomes the Baseline Programme.

Buffer Activity: an activity in a programme that acts as a Contingency or to artificially absorb Float.

Calculate Schedule: the mathematical analysis of a network, usually using a computer and Project Management Software, to determine the earliest and latest starts and finishes and float of the activities and the overall project duration. Often carried out following the addition of actual progress to determine the effect of progress on the network; primarily the completion date of the project.

Change or Variation: any difference between the circumstances and/or content of the contract works as carried out, compared with the circumstances and/or content under which the works are described in the contract documents as required to be or intended to have been carried out. A change or variation may or may not carry with it a right to an extension of time and/or additional payment.

Collapsed As-Built: a method of delay analysis where the effects of events are 'subtracted' from the As-Built Programme to determine what would have occurred but for those events.

Compensation: the recovery or payment of money for work done or time taken up whether by way of valuation, loss and/or expense or damages.

Compensable Event: expression sometimes used to describe an Employer Risk Event in respect of which the Contractor is entitled to compensation.

Concurrency: true concurrent delay is the occurrence of two or more delay events at the same time, one an Employer Risk Event, the other a Contractor Risk Event and the effects of which are felt at the same time. The term 'concurrent delay' is often used to describe the situation where two or more delay events arise at different times, but the effects of them are felt (in whole or in part) at the same time. To avoid confusion, this is more correctly termed the 'concurrent effect' of sequential delay events.

Constraints: Restrictions that affect the sequence or timing of an activity. These include Predecessor dependencies but more often refer to Imposed Dates.

Constructive Acceleration: acceleration following failure by the Employer to recognise that the Contractor has encountered Employer Delay for which it is entitled to an EOT and which failure required the Contractor to accelerate its progress in order to complete the works by the prevailing contract completion date. This situation may be brought about by the Employer's denial of a valid request for an EOT.

Contract Administrator (CA): the person responsible for administration of the contract, including certifying what extensions of time are due, or what additional costs or loss and expense are to be compensated. Depending on the form of contract the person may be referred to by such terms as Employer's Agent, Employer's Representative, Contract Administrator, Project Manager or Supervising Officer or be specified as a particular professional, such as the Architect or the Engineer. The contract administrator may be one of the Employer's employees.

Contract Completion Date: the date by which the Contractor is contractually obliged to complete the works. As well as being an overall date for completion, the contract completion date may be the date for completion of a section of the works or a milestone date. The expression 'completion date' is sometime used by Contractors to describe the date when they plan to complete the works (which may be earlier than the contract completion date).

Contractor Risk Event: an event or cause of delay which under the contract is at the risk and responsibility of the Contractor.

Critical Activity: an activity with Zero Float. If a critical activity is delayed or extended, it will delay or extend the completion of the project and, generally, if a critical activity is advanced or reduced it will advance or reduce the completion of a project.

Critical Delay: a delay to progress of any activity on the critical path will, without acceleration or resequencing, cause the overall project duration to be extended, and is therefore referred to as a 'critical delay'.

Critical Path: the sequence of activities through a project network from start to finish, the sum of whose durations determines the overall project duration. There may be more than one critical path depending on workflow logic.

Critical Path Analysis (CPA) and Critical Path Method (CPM): the critical path analysis or method is the process of deducing the critical activities in a programme by tracing the logical sequence of tasks that directly affect the date of project completion. It is a methodology or management technique that determines a project's critical path. The resulting programme may be depicted in a number of different forms, including a Gantt or barchart, line-of-balance diagram, pure logic diagram, time-scaled logic diagram or as a time-chainage diagram, depending on the nature of the works represented in the programme.

Culpable Delay: an expression sometimes used to describe a Contractor Delay.

Delay Event: an event or cause of delay which may be either an Employer Risk Event or a Contractor Risk Event.

Delay to Completion: this expression may mean either delay to the date when the contractor planned to complete its works, or a delay to the contract Completion Date.

Delay to Progress: this means a delay which will merely cause delay to the contractor's progress without causing a contract Completion Date not to be met. It is either an Employer Delay to Progress or a Contractor Delay to Progress.

Delay Analysis: The methodological investigation of the causes and effects of activities, or sequences of activities, completing later than planned.

Dependency: Logical interrelationships between activities. In a network there can be one or more dependency between any two activities. There are four types of dependency; 'finish to finish', 'finish to start', 'start to finish' and 'start to start'. See also Predecessor, Successor. The dependencies dictate the sequence in which activities can be carried out.

Disruption: disturbance, hindrance or interruption of a contractor's normal work progress, resulting in lower efficiency or lower productivity than would otherwise be achieved. Disruption does not necessarily result in a Delay to Progress or Delay to Completion.

Driving: where there are a number of dependencies to an activity the driving dependency is the dependency, or dependencies, that result in the timing of the start and/or finish of that activity. The implication is that there is Free Float on non-driving dependencies.

Duration: is the length of time needed to complete an activity. The time period can be determined inductively by determining the start and finish date of an activity, or deductively by calculation from the time necessary to expend the resources applied to the activity.

Early Finish: of an activity, is the earliest programmed calendar date on which an activity can be finished before any of its succeeding float is consumed.

Early Start: of an activity, is the earliest programmed calendar date on which an activity can be started before any of its preceding float has been consumed.

Employer Delay: expression commonly used to describe any delay caused by an Employer Risk Event.

Employer Delay to Completion: a delay which will cause a Contract Completion Date not to be met.

Employer Delay to Progress: a delay which will merely cause delay to the contractor's progress without causing a Contract Completion Date not to be met.

Employer Risk Event: an event or cause of delay which under the contract is at the risk and responsibility of the Employer.

Excusable Delay: expression sometimes used to describe an Employer Delay in respect of which the contractor is entitled to an Extension of Time.

Extension of Time (EOT): additional time granted to the contractor to provide an extended contractual time period or date by which work is to be, or should be completed and to relieve it from liability for damages for delay (usually liquidated damages).

Float or Slack: the amount of time between the early start date and the late start date, or the early finish date and the late finish date, of any of the activities in a programme.

Free Float: is the amount of time an activity can be delayed beyond its early start/early finish dates without delaying the early start or early finish of any immediately following activity.

Global Claim: a global claim is one in which the Contractor seeks compensation for a group of Employer Risk Events but does not or cannot demonstrate a direct link between the loss incurred and the individual Employer Risk Events.

Hammock: an activity representing the period from the start of an activity to the completion of another. Sometimes used as a way of summarising the duration of a number of activities in a programme as one single duration.

Hanging Activity: an activity not linked to any preceding or successor activities. Sometimes called a 'dangling activity'.

Impact: the effect that a change has on an activity or the effect that a change to one activity has on another activity.

Key Date: expression sometimes used to describe a date by which an identifiable accomplishment must be started or finished. Examples include 'power on', 'weathertight' or the start or completion of phases of construction or of phases or sections of the contract, or completion of the works.

Lag: lag in a network diagram is the minimum necessary lapse of time between the finish of one activity and the finish of another overlapping activity. It may also be described as the amount of time required between the start or finish of a predecessor task and the start or finish of a successor task. See also Logic Links.

Lead: the opposite of lag, but in practice having the same meaning. A preceding activity may have a lag to a successor activity – from the perspective of the successor activity, that is a lead.

Liquidated and Ascertained Damages (LADs) or Liquidated Damages (LDs): a fixed sum, usually per week or per day, written into the contract as being payable by the contractor in the event that the works are not completed by the contract Completion Date, original or extended.

Logic Links: the normal links are as follows:
  Finish-to-start (FS)
  Lagged finish-to-start (FS+/−)
  Start-to-start (SS)
  Lagged start-to-start (SS+/−)
  Finish-to-finish (FF)
  Lagged finish-to-finish (FF+/−)
  Lagged start and finish (SF+/−)
Method Statement: a written description of the contractor's proposed manner of carrying out the works or parts thereof, setting out the assumptions underlying the programme, the reasoning behind the approach to the various phases of construction and listing all the work encapsulated in the programme activities. It may also contain the activity duration calculations and details of key resources and gang strengths.
Milestone: a key event selected for its importance in the project. Commonly used in relation to progress, a milestone is often used to signify a key date.
Mitigation: mitigate means making less severe or less serious. In connection with Delay to Progress or Delay to Completion, it means minimising the impact of the Risk Event. In relation to disruption or inefficient working, it means minimising the disruption or inefficiency. Failure to mitigate is commonly pleaded as a defence or partial defence to a claim.
Must Start/Must Finish: most project management software allows the user to specify that an activity must start or must finish on a specific date. Using the software in this way restricts the ability of the programme to react dynamically to change on the project.
Negative Lag: see Logic Links above.
Negative Total Float: expression sometimes used to describe the time by which the duration of an activity or path has to be reduced in order to permit a limiting imposed date to be achieved. Negative float only occurs when an activity on the critical path is behind programme. It is a programming concept, the manifestation of which is, of course, delay.
Non-compensable Event: expression sometimes used to describe what the Protocol calls a Contractor Risk Event.
Non-excusable Delay: expression sometimes used to describe what the Protocol calls Contractor Delay.
PERT (Programme Evaluation and Review Technique): a programming technique, similar to critical path analysis, but whereby the probability of completing by the contract completion date is determined and monitored by way of a quantified risk assessment based on optimistic, pessimistic and most likely activity durations.
Planned Completion Date: see Contract Completion Date.
Practical Completion: the completion of all the construction work that has to be done, subject only to very minor items of work left incomplete. It is

generally the date when the obligation to insure passes from the Contractor to the Employer and the date from which the defects liability period runs. This is the term used under the Joint Contracts Tribunal (JCT) family of contracts. Under the Institution of Civil Engineers (ICE) forms and in the International Federation of Consulting Engineers (FIDIC) forms it is referred to as Substantial Completion.

Precedence Diagram: a multiple dependency, activity-on-node network in which a sequence arrow represents one of four forms of precedence relationship, depending on the positioning of the head and the tail of the sequence arrow. See Logic Links.

Programme: the programme illustrates the major sequencing and phasing requirements of the project. Otherwise known as the Schedule.

Prolongation: the extended duration of the works during which costs are incurred as a result of a delay.

Resource: expression used to describe any variable capable of definition that is required for the completion of an activity and may constrain the project. This may be a person, item of equipment, service or material that is used in accomplishing a project task.

Resource Levelling: expression used to describe the process of amending a schedule to reduce the variation between maximum and minimum values of resource requirements. The process removes peaks, troughs and conflicts in resource demands by moving activities within their early and late dates and taking up float. Most project planning software offers an automated resource-levelling routine that will defer the performance of a task within the imposed logical constraints until the resources assigned to the tasks are available.

Risk Event: see Employer Risk Event and Contractor Risk Event.

Substantial Completion: see Practical Completion.

Time Impact Analysis: method of delay analysis where the impacts of particular delays are mapped out at the point in time at which they occur, allowing the discrete effect of individual events to be determined.

Total Float: the amount of time that an activity may be delayed beyond its early start/early finish dates without delaying the contract completion date.

Updated Programme: in the Protocol the Updated Programme is the Accepted Programme updated with all progress achieved. The final Updated Programme should depict the as-built programme.

Works: what the Contractor is obliged to construct is referred to as the Works.

Appendix 2
Standards for the Levels of a Programme or Schedule

Client's or owner's programme/schedule

The purpose of the client's or owner's programme/schedule is to set out the clients/owners desired overall time requirements for the project.

A number of different types of organisation might produce the client's/owner's programme or schedule. This could be the client/owner themselves, their project manager, an architect, engineer or QS. This will vary from industry to industry and will be largely dependent upon the structure and contractual arrangements of the individual project.

The purpose of this programme/schedule is to inform the project participants what the client/owner wants to be achieved. This programme/schedule is sometimes used as the basis for obtaining tenders for the work that follows. It is also sometimes used to describe the client's vision or aspirations, in terms of time, appointments, phasing, handovers and the like.

Rarely is this programme/schedule based on a critical path analysis. Sometimes, it is not presented as a programme or schedule at all, instead being described by a set of dates in a contract, enquiry document, or as a set of milestone target dates to be achieved. This often lays out the major date time frames for design, procurement, assembly, production, construction, etc.

This programme/schedule has not been included within the standard levels as defined in this section as this programme/schedule invariably is not part of the main planning/scheduling undertaken on the project – it defines the key stages and dates of all of the work that follows.

Practical Guide to Disruption and Productivity Loss on Construction and Engineering Projects, First Edition. Roger Gibson.
© 2015 John Wiley & Sons, Ltd. Published 2015 by John Wiley & Sons, Ltd.

2 Proposed levels of programme/schedule for a single project

At the end of this section, in Attachment 1, is the proposed standardised levels of programme/schedule for a single project. Later, the levels for an entire range of projects, or programme of projects, are described.

The various programmes or schedules produced for a single project are generally produced using five levels, and each is discussed within this section.

The basic premise of this paper is that programmes/schedules should be produced using a standard set of levels:

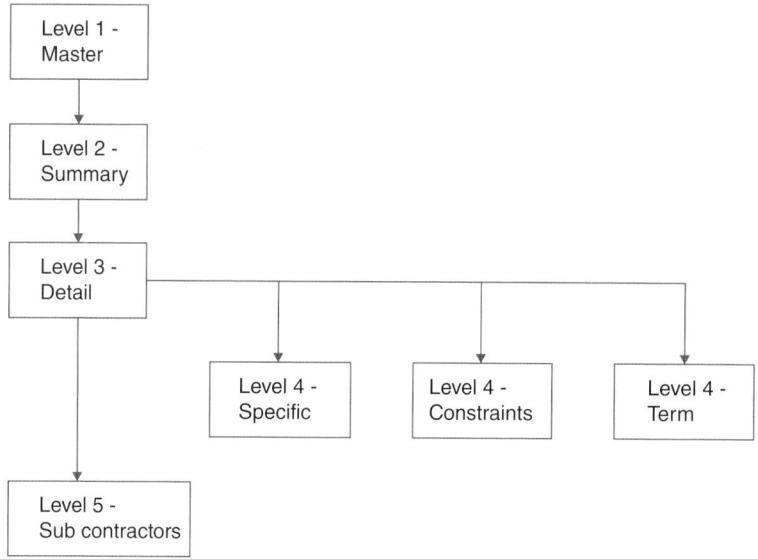

Level 1: Overall project master programme/schedule for the project

The purpose of the level 1 programme/schedule is to show the overall timing of all aspects of the project.

This programme/schedule will show the overall coordinated timing of all the key stages of the project. This will include topics such as feasibility, design, procurement, manufacture, assembly, production, construction, installation and commissioning.

Sometimes, this level 1 programme/schedule is used during the initiation or feasibility stage of a project and may be used as part of an overall business plan.

This is the high-level programme/schedule against which the overall timing of the project is set out and communicated. It is also the programme/schedule against which the overall progress of the project is reported to the client/owner. Often, this programme/schedule is presented not only as a barchart, but also with a set of milestone dates against which the project is monitored.

Ideally, this programme/schedule should be no more than a single sheet, containing perhaps 30 to 100 activities (dependent upon the complexity of the project). The target milestone dates defined by the client/owner are often included in this level 1 programme/schedule. This programme/schedule will normally illustrate the critical path of the project.

The important point here is to consider who will receive and review this level 1 programme/schedule. Very often, it is looked at by people who do not readily understand programmes and schedules and therefore the complexity of the level 1 programme/schedule should take this into account. There is little point in producing a level 1 programme/schedule that cannot be understood by those who receive it.

Level 2: The summary level programmes/schedules for the project

The main purpose of these programmes/schedules at level 2 is to set out when each of the key elements will take place. For example, when each part of the design will be undertaken, or each part of the procurement/assembly, or each part of the installation, production or construction. These programmes/schedules illustrate a summary of the more detailed levels that follow.

These level 2 summary programmes/schedules will often be a suite of linked individual programmes for each of the key elements of the project, such as design, procurement and production or construction. Alternatively, this could be one programme that includes the work of all key elements of the project, with the programmes of individual elements, such as design, procurement, production or construction, being created by using filters to select which activities are to be represented on individual programmes/schedules.

Ideally, these various programmes/schedules will all be produced or coordinated by a single party, although this depend upon the contractual arrangements for the project.

The entire scope of the project should be covered at this and subsequent levels. These summary programmes will show the critical path for the project. Another main purpose may be the control of the project by EVA, CPA, etc.

These summary programmes should be in sufficient detail to do three things:

1. To enable all involved in that element to fully understand what needs to be done when and by whom.
2. To enable those involved in a particular element to fully understand how their own work interfaces with that work in another element (or within the same element). For the purposes of this section, different elements might be design, procurement, installation, production or construction. A good example of this is the design of the project. The level 2 summary design programme will be used by, say, the procurement or construction team to understand how the timing of the design will affect the procurement or installation/construction activities that follow.

3. To enable the project progress to be monitored and progress reported at this summary level.

The number of activities on these level 2 programmes will obviously depend on the complexity of the project. As a suggestion, individual activities should be no longer than 4–6 weeks each, as longer periods become difficult to assess progress. In most cases, it will be essential for each of the level 2 programmes to be logically linked together, in order that the implications of progress achieved on one element can be seen on another element.

On smaller projects, these level 2 summary programmes/schedules may not be needed.

Level 3: The detailed level

The purpose of these detailed level 3 programmes/schedules is to show the in-depth timing of all of the activities on the project. These will definitely show the critical path, and will often include details of the resources needed to undertake that element or stage or the work.

Sometimes, depending on the contractual arrangements for the project, one party will be responsible for producing all of these level 3 programmes/schedules and will also be responsible for the coordination between them. For example, this will almost certainly be the case on a project being undertaken on a design-and-build contractual arrangement.

In other instances, say where the design is undertaken by one party and the installation/production/construction by a different separate party (for example, in UK traditional fixed price construction contracts), these various level 3 programmes/schedules might be produced by different organisations.

Examples of the different types of level 3 programme/schedule could include the following:

- The detailed design programme would show every drawing to be produced.
- The detailed authorities' approvals and statutory processes programme.
- The detailed procurement programme would show the timing of subcontract tenders to be obtained, placing orders, manufacture times, lead times, delivery dates and the like.
- The detailed production, construction, installation or assembly programmes/schedules would show the timing of all physical works on site, for each part of the project.

Additionally, each of the programmes/schedules could be further broken down into off-site and onsite works.

Often, these level 3 detailed programmes are produced in advance of appointing the various specialists (designers and subcontractors) in order to set out the overall detailed timing of activities and interfaces on the project. This is often difficult to achieve as the input from those specialists is required to fully test the detail. Also, some degree of replacing the original detail will be required once the various specialists' own programmes/schedules are received.

Depending upon the contractual arrangements for the project, these programmes/schedules might or might not be shared with the client. Whether they are or not, their purpose is to clearly set out when each individual detailed part of the project is to be undertaken and by whom.

Level 4: The specific, term and constraints level

The main purpose of this level is to extract from level 3 key information that is then used to create further programmes/schedules at level 4. These are a group of programmes/schedules often produced by filtering the detail of the level 3 programmes although sometimes these level 4 programmes are standalone programmes not linked electronically to other levels.

These programmes/schedules cover a wide range of uses, ttypically:

- Programmes/schedules that are produced to cover specific areas or aspects of the project.
- Short- and medium-term programmes/schedules covering the detailed activities for the next few days, weeks or months. These are sometimes referred to as look-ahead programmes/schedules.
- Programmes/schedules specifically produced to give guidance to individual subcontractors about when their work will need to be carried out. These are sometimes called constraints programmes.

Often these specific level programmes/schedules are derivatives of the level 3 detailed schedules/programmes and are created by applying a number of different filters based on the coding structure and capabilities of the software used. These filters/coding structures can be used, for example, to produce programmes/schedules of work for each manager, each subcontractor, or for different areas of work or location, by floor level, etc.

Level 5: The subcontractors' programmes/schedules

The purpose of these level 5 programmes/schedules is for the subcontractors to set out the detailed timing of their own work. These programmes/schedules are often produced by subcontractors to further define their work if the detail contained at level 3 is insufficient for their purposes. No matter what system the subcontractor uses, the overall timeframe for their work must conform to the timeframes set out in the level 3 programme/schedule. The subcontractor uses these programmes/schedules to monitor and report progress of their own works.

Most often, these programmes/schedules are held in different software databases than the level 1, 2, 3 and 4 programmes/schedules.

The particular contractual arrangements and processes involved might determine that the level 5 programmes are not electronically linked to the levels above. They might be regarded as stand-alone in their own right. However, whether electronically linked to the levels above or not, it is essential that the critical path for each individual subcontractor's work is shown on their own programme. Later, if there is a claim from a subcontractor, it is crucial that the subcontractors' critical path originally anticipated was shown in order to assess the validity of that claim.

There are essentially two approaches to the subcontractors' programmes/schedules. The subcontractors could either produce these as stand-alone programmes/schedules, or they could help to develop the level 3 programmes/schedules, with their own specific activities being included within the main level 3 programme/schedule.

3 Smaller Projects

On some smaller projects different arrangements may be used for the various levels of programming/scheduling.

Depending upon the size of the project, levels 3, 4 and 5 are sometimes combined together into one single level. This can incorporate the work of all parties into one detailed programme, with one party retaining responsibility for that detailed programme.

Additionally, on some smaller projects, the level 2 summary programmes/schedules may not be necessary.

4 Proposed Levels of Programme/Schedule for a Group or Programme of Projects

At the end of this section, Attachment 2 shows the proposed standardised levels of programme/schedule for a group or programme of projects.

This is almost exactly the same as described for an individual project in section 3 above, in that the levels of programme/schedule for the each project within a group or programme of projects will be the same as that described in section 2.

The main difference with a group or programme of projects is that an additional level of programme/schedule, called level 0, is introduced to show the summary of all projects within the programme.

The level of detail in this level 0 programme/schedule depends upon how many projects make up that group or programme. If it is a small number, say 3 or 4, then each of the individual projects could be represented by, say 10 activities. If there are a large number of individual projects making up the group or programme, then each project might be represented by only one activity.

5 Important Associated Considerations

When reviewing the various programmes/schedules defined in this section, one also has to consider how they will be created. There could be two approaches to this.

The first is whether all of the various levels are created as one integrated programme/schedule. One can see that the various levels could be produced working towards what will eventually be a single, totally integrated, fully logic linked programme/schedule that is divided into the various levels. This can be achieved by creating the level 1 programme/schedule first, with level 2 being a detailed expansion of level 1, the level 2 programme/schedule is then in turn further detailed to create level 3, and so on. However, in other organisations, the detailed levels of programme/schedule are created first and this generates the master level.

This approach ensures that all activities are fully integrated with all other activities at every level. This is without doubt the most suitable option as it aids simplicity – having all of the programme/schedule information in one place maintains control and limits the chance of confusion or, even worse, uncoordinated programmes/schedules with the inevitable problems that will arise.

The second approach is potentially more difficult, in that the various levels are produced as individual programmes/schedules that are not integrated but maintained as stand-alone items. This approach is sometimes practically easier to achieve, in that the various parties responsible for certain aspects of the scheduling – say a design firm, a manufacturing firm, a construction contractor and subcontractors – will find it easier to produce their own programmes/schedules. However, this approach can lead to the content being held in different locations on different databases and thus lead to confusion and out-of-date information being used.

Whichever approach is used, what is important is that once the various levels of programme/schedule have been completed and approved by all, they become the as-planned programme/schedule against which all subsequent programmes/schedules are measured.

Occasionally, one hears of projects that run two sets of programmes/schedules – the master set and what is commonly known as a target programme or schedule. The latter is often used in an attempt to shorten the overall duration of the project, but this method is fraught with difficulties. The is great potential for confusion as to what programme/schedule different organisations, designers, contractors and sub-contractors are working to. Extreme caution must be exercised if this approach is adopted.

Additionally, when creating the various levels of a programme/schedule one often also has to consider the impact of the Work Breakdown Structure (WBS) of the project. From a programming/scheduling point of view, there is often no real requirement to take the WBS into consideration, but practical application shows that this is helpful on complex projects in order to find common ground between cost/budget and programming/scheduling.

184 *Appendix 2*

The various programmes produced at prequalification and tender stage have deliberately been excluded from this section. This is due to the fact that, most often, these are separate programmes/schedules which are not linked to the main project programmes/schedules.

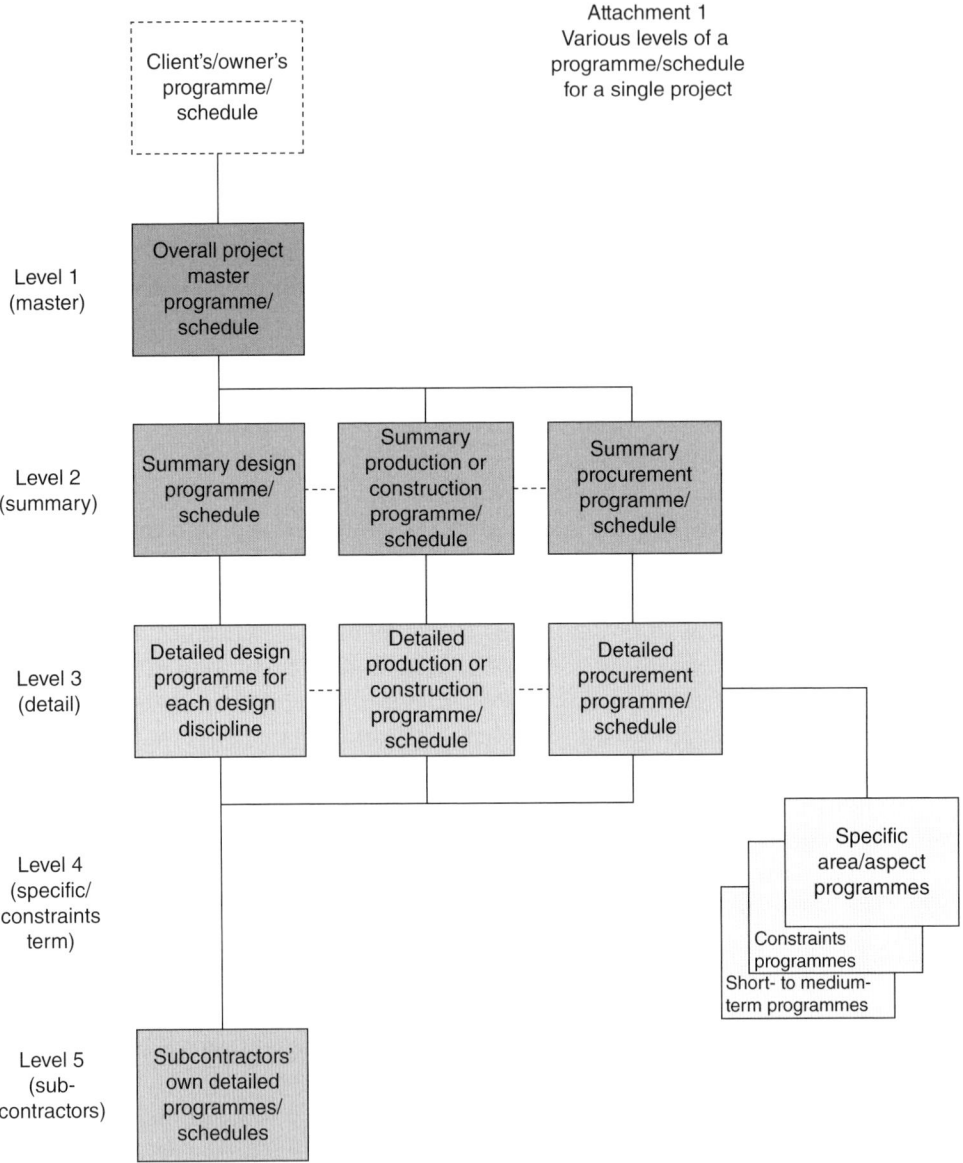

Attachment 1
Various levels of a programme/schedule for a single project

Appendix 2

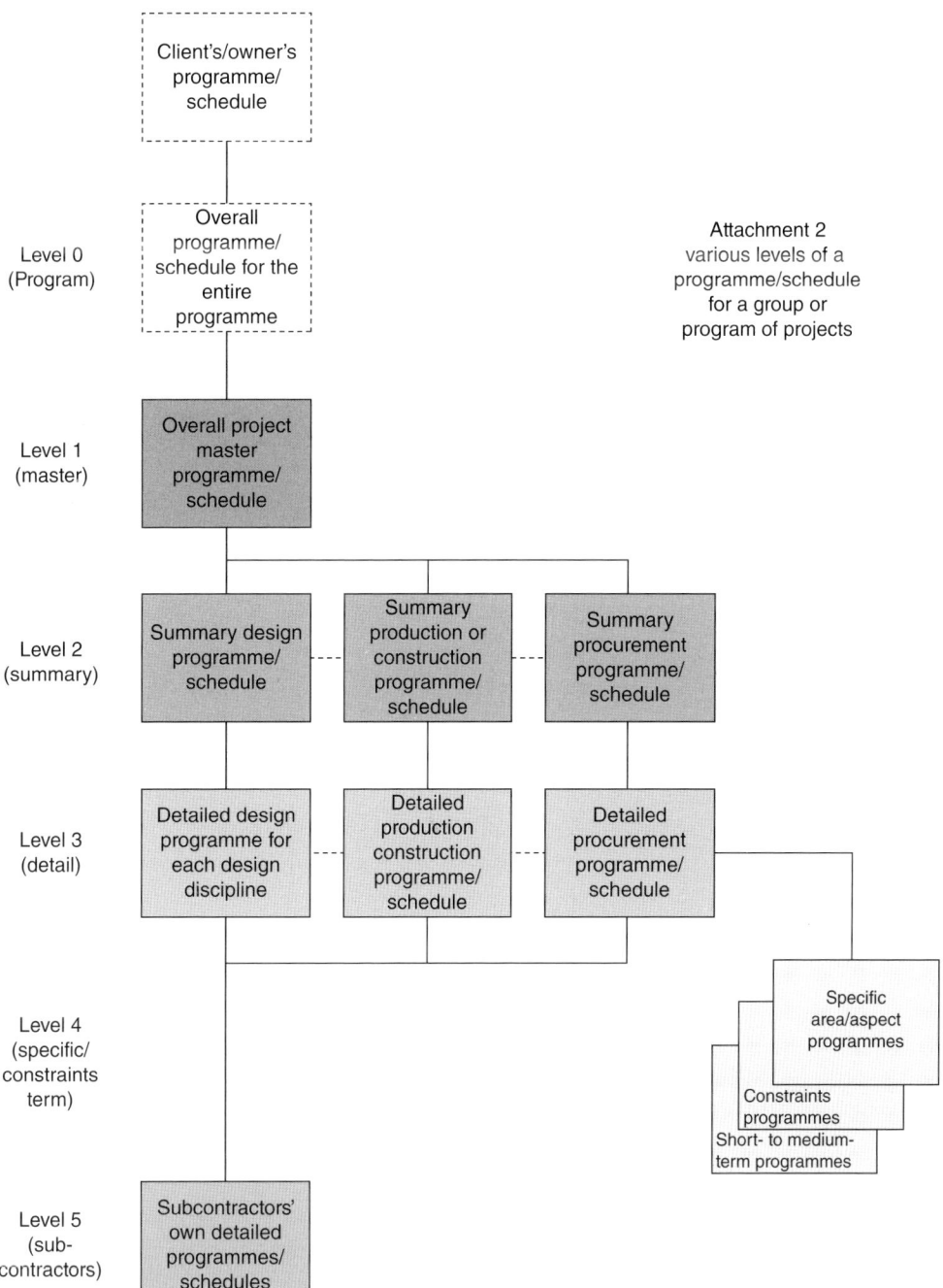

Appendix 3
Society of Construction Law: Delay & Disruption Protocol (October 2002)

Guidance Clauses

Clause 19: Disruption

1.19.1 Disruption (as distinct from delay) is disturbance, hindrance or interruption to a Contractor's normal working methods, resulting in lower efficiency. If caused by the Employer, it may give rise to a right to compensation either under the contract or as a breach of contract.

Guidance

1.19.2 Disruption is often treated by the construction industry as if it were the same thing as delay. It is commonly spoken of together with delay, as in 'delay and disruption'. Delay and disruption are two separate things. They have their normal everyday meanings. Delay is lateness (e.g. delayed completion equals late completion). Disruption is loss of productivity, disturbance, hindrance or interruption to progress. In the construction context, disrupted work is work that is carried out less efficiently than it would have been, had it not been for the cause of the disruption.

1.19.3 Disruption to construction work may lead to late completion of the work, but not necessarily so. It is possible for work to be disrupted and for the contract still to finish by the contract completion date. In this situation, the Contractor will not have a claim for an EOT, but it may have a claim for the cost of the reduced efficiency of its workforce.

1.19.4 Not all disruption is subject to the payment of compensation. The Contractor will be able to recover disruption compensation only to the extent that the Employer causes the disruption. Most standard forms of contract do not deal expressly with disruption. If they do not, then

---

Practical Guide to Disruption and Productivity Loss on Construction and Engineering Projects, First Edition. Roger Gibson.
© 2015 John Wiley & Sons, Ltd. Published 2015 by John Wiley & Sons, Ltd.

disruption may be claimed as being a breach of the term generally implied into construction contracts, namely that the Employer will not prevent or hinder the Contractor in the execution of its work.

1.19.5 Disruption has to be established in the normal cause and effect manner. Disruption is not just the difference between what actually happened and what the Contractor planned to happen. The objective of the compensation for disruption is to put the Contractor in the same financial position it would have been in if the disruption had not occurred.

1.19.6 The most common causes of disruption are loss of job rhythm (caused by, for example, premature moves between activities, out of sequence working and repeated learning cycles), work area congestion caused by stacking of trades, increase in size of gangs and increase in length or number of shifts. But these are also symptoms of poor site management.

1.19.7 The starting point for any disruption analysis is to understand what work was carried out, when it was carried out and what resources were used. For this reason, record keeping is just as important for disruption analysis as it is for delay analysis. The most appropriate way to establish disruption is to apply a technique known as 'the Measured Mile'. This compares the productivity achieved on an un-impacted part of the contract with that achieved on the impacted part. Such a comparison factors out issues concerning unrealistic programmes and inefficient working. The comparison can be made on the man-hours expended or the units of work performed. However care must be exercised to compare like with like. For example, it would not be correct to compare work carried out in the learning curve part of an operation with work executed after that period.

1.19.8 It may be difficult to find un-impacted parts on some contracts. Comparison of productivity on other contracts executed by the Contractor may be an acceptable alternative, provided that sufficient records from the other contracts are available to ensure that the comparison is on a like for like basis. Failing that, it might be acceptable to use model productivity curves and factors developed by a number of organisations from data collected on a range of projects (e.g. by the US Army Corps of Engineers, International Labour Organisation, Mechanical Contractors' Association of America Inc., Chartered Institute of Building, etc.). These curves provide general guidance and they should be used only if they are relevant to the working conditions and type of construction and are supported by evidence from the party seeking to prove disruption.

1.19.9 When establishing the compensation for disruption it is necessary to isolate issues that can affect productivity but are unrelated to the Employer's liability. These issues can include weather, plant breakdowns, dilution of supervision, contractor management and acceleration. The Contractor has an obligation to manage its own change efficiently and any failure to do this should not be compensated.

1.19.10 The use of unsupported percentage additions to assess disruption is not advocated. However, on very simple contracts where evidence of disruption can be demonstrated, this practice may be acceptable, provided that the percentage addition is small (for guidance, no more than 5% on labour and plant).

1.19.11 It is essential for the Contractor to maintain and make available to the CA good site records in order that the CA may carry out proper assessments of disruption. The Contractor should also be required to give prompt notice of disruption, in order that the CA can promptly investigate the claim.

1.19.12 It is recommended that the compensation for disruption caused by variations should be agreed as soon as possible after completion of the variation and where practicable included with the valuation of the variation (see Guidance Section 1.7.7).

1.19.13 It is recommended that the compensation for disruption caused by other events that are the liability of the Employer be compensated by the actual reasonable costs incurred, plus a reasonable allowance for profit if allowed by the contract.

Index

absenteeism, 129, 130
accelerate(s), 21, 22, 54, 134, 135, 153, 154, 156, 157, 159–164, 171
accelerated, 135, 142, 154, 158, 159, 163
accelerating, 7, 48, 56, 113, 132, 162
acceleration, 3–5, 7, 11–13, 18, 19, 21, 22, 28, 32, 41, 42, 45–56, 59, 100, 101, 103, 112, 113, 120, 129, 134, 135, 139, 142, 149, 153–67, 169, 171, 188
access, 35, 38, 71, 79, 80, 86, 127, 131, 132
activity(ies), 7, 8, 15, 35, 44, 62–71, 75, 77, 81–4, 88–92, 95–7, 103, 104, 108, 113–15, 117, 119, 122, 126–8, 132, 134, 140, 141, 158, 163, 166, 169–75, 179–83, 188
allowances, 5, 7, 80–82, 107
amendment, 21, 79, 85
analysis, 5, 9, 24, 26, 27, 32, 37, 41–7, 58, 70, 71, 80, 99, 100, 103–7, 109, 117, 119–21, 125, 128, 133–9, 145, 146, 152, 162, 170–172, 174, 175, 177, 188
arbitration, 1, 100, 157
arbitrator(s), 1, 4, 99, 105, 107, 112, 157
architect, 7–9, 11, 13–17, 79, 85, 113, 128, 134, 153, 155–8, 166, 167, 171, 177
assessment(s), 37, 39, 43, 49, 51, 54, 58, 81, 84, 87, 109, 111, 138, 142, 148–51, 174, 189
attributable, 41, 43, 44, 46, 47, 49–52, 55–9, 102

baseline, 68, 78, 82, 84, 88, 89, 92, 119, 120, 124, 132, 133, 140, 142, 170
baselined, 137

causation, 24, 25, 99–123, 126, 133, 140, 142, 152, 157

causative, 24, 37, 50, 54, 57, 106, 109, 119
causatively, 37, 39, 42, 47
claim(s), 1–3, 5–7, 9, 11, 13, 19–34, 36–9, 42, 44, 45, 48, 50–53, 84, 90, 91, 95–7, 99–101, 103, 105–13, 116–20, 123, 124, 126, 133–6, 140, 152, 153, 156–9, 161–3, 165, 166, 173, 174, 182, 187, 189
claimant(s), 23, 33, 107, 108, 110, 118, 166
compensated, 126, 142, 171, 188, 189
compensation, 4–7, 80, 81, 85, 87, 99, 100, 102, 105–108, 112, 116, 117, 120, 126, 158, 165, 166, 170, 173, 187–9
concurrency, 4, 6, 102, 171
concurrent, 4, 6, 17, 99, 101–3, 110, 121, 171
concurrently, 17, 109, 129
constraint(s), 7, 68, 71, 81–3, 86, 90, 91, 113, 169–71, 175, 181
constructive, 5, 100, 112, 129, 139, 153, 154, 156–60, 171
contemporaneous, 8, 41, 43, 58, 69, 97, 98, 106, 108, 110–111, 114, 126, 134, 139, 141, 144, 145, 147
contemporaneously, 88, 123, 126
cost(s), 1, 2, 4–7, 18, 19, 22, 23, 26, 28–37, 41–5, 47–59, 91, 97, 100–102, 106–9, 113, 116, 117, 119, 123, 125–30, 133, 134, 136–40, 142, 150–152, 154–64, 166, 167, 169–171, 175, 183, 187, 189
counterclaim(s), 20, 28–31, 33, 34, 41, 44–7, 55–7
counterclaimed, 44
critical, 2, 7, 38, 42, 45, 47, 48, 63, 64, 66–70, 75, 83, 88, 90–92, 96, 103, 108, 113, 120–122, 137, 161, 171, 172, 174, 177, 179, 180, 182
criticality, 37, 39
critically, 123, 125

damages, 1, 6, 7, 20, 22, 28–30, 44, 51, 55, 56, 102, 118, 126, 129, 135, 136, 142, 156, 157, 159, 161, 164–6, 170, 173
decision, 4, 14, 17, 18, 22, 30, 44, 75, 79, 85, 99, 100, 107, 112, 120, 137, 157, 159
decline, 128–32, 155, 158
defects, 41, 44, 45, 85, 175
defined, 14, 16, 20, 25, 27, 89, 100, 105, 154, 159, 177, 179, 183
delay(s), 1–9, 11, 13, 14, 16, 17, 19, 21, 27, 29, 30, 33–9, 41–53, 55–9, 64, 66, 67, 70, 71, 78, 83–5, 87, 90, 92, 96, 97, 99–122, 129, 132–5, 138, 142, 149, 153, 154, 156–61, 165, 166, 169–75, 187–9
delayed, 11, 14, 17, 29, 35, 42, 46, 66–8, 71, 75, 81, 88, 104, 106, 132, 140, 141, 143, 145, 153, 161, 165, 171, 173, 175, 187
delaying, 8, 42, 47, 95, 102, 104, 121, 145, 173, 175
demonstrate, 2, 3, 35–9, 97, 104, 108, 111, 117, 129, 138, 142, 152, 156, 163, 166, 173
demonstrated, 31, 35, 37, 105, 114, 163, 189
dependent, 68, 87, 115, 177, 179
disallow, 54
dispute(s), 1, 2, 4, 5, 8, 11–13, 18–20, 28, 30, 32, 38, 57, 65, 84, 98–100, 107, 109, 110, 112, 114, 153, 154, 162, 163
disputed, 5, 9, 20
disputing, 25
disrupting, 116
disruption(s), 1–9, 11–13, 18–20, 23, 24, 26–8, 30–39, 41–5, 47, 55, 57, 58, 61, 96, 98–123, 134, 138, 139, 141, 142, 147–9, 153, 154, 165, 169, 172, 174, 177, 187–9
disruptive, 2, 37, 98, 108, 115, 135
disturbance, 5, 15, 108, 112, 172, 187
downtime, 41–3, 46, 47, 49, 58
driving, 69, 121, 172
duration(s), 8, 46, 63, 66, 68–71, 75, 82–4, 87, 89, 101, 114, 122, 132, 158, 160, 169–175, 183

earned, 41, 42, 58, 92, 115, 116, 124, 126–8, 133, 137–9, 144, 146, 147, 150–152
effect, 2, 4–6, 19, 29, 38, 39, 42, 45, 47, 49, 53, 62–4, 67, 68, 70, 75, 80, 81, 85, 87, 97, 100, 101, 104–7, 110–112, 116, 119–22, 126, 128–30, 133, 136, 149, 158, 160, 165–7, 170, 171, 173, 175, 188
efficiency(ies), 2, 5, 108, 112, 118, 123–33, 136, 138–40, 142–4, 146–9, 151, 166, 172, 187

efficient, 34, 129, 132, 133, 138, 163
efficiently, 130, 156, 166, 187, 188
effort(s), 42, 47, 52–6, 61, 109, 123, 125, 133, 140, 159, 160, 162, 163, 166
endeavours, 14, 154, 156, 165, 166
entitled, 4, 6–8, 21, 22, 32, 35, 39, 44, 49, 52, 100, 102, 113, 117, 118, 134, 154, 156–9, 161, 167, 170, 171, 173
entitlement, 4–6, 22, 30, 34, 44, 51, 55, 57, 97, 99, 103–5, 108, 112, 123, 138, 142, 153, 154
entitles, 118
evaluates, 49
evaluating, 1, 2, 42, 58, 67
evaluation, 2, 4, 5, 7, 44, 63, 105, 174
event(s), 2, 4, 5, 8, 9, 11, 14–15, 17, 24, 29, 34, 35, 37–9, 42, 46–8, 53, 80, 81, 84, 85, 87, 90, 98, 99, 101–112, 115–17, 119–22, 128, 129, 133, 137, 138, 140, 141, 145, 147, 152, 157, 158, 165–7, 169–75, 189
evidence, 2, 8, 20, 24, 28, 31, 32, 34–9, 41–4, 46–8, 50, 51, 53–6, 58, 103, 105, 106, 114, 119, 188, 189
evidenced, 19
evidencing, 107
excusable, 99, 101, 102, 154, 159, 160, 173
expected, 17, 63, 89, 105, 110, 116, 129, 131, 137, 161
expend, 166, 172
expended, 114, 126, 127, 133, 139, 143–5, 146, 188
expending, 123, 125
expenditure, 14, 16, 28, 31, 128
expense, 1–3, 5, 7, 11, 13, 15, 16, 33, 34, 36–9, 100, 103, 106, 155–8, 166, 170, 171
expert(s), 3, 24, 30–32, 36, 38, 41, 45, 47, 49, 54, 58, 59, 99, 104, 106, 107, 112, 120, 135, 136, 139
extend, 4, 112, 165, 171
extended, 5, 45–7, 48, 50, 56, 57, 100, 105, 132, 155, 157, 171, 173, 175
extension(s), 1, 3–9, 11, 13, 14, 17, 22, 28, 78, 84, 85, 90, 97, 99, 101–5, 107, 112, 135, 153–61, 166, 170, 171, 173

fact(s), 9, 18, 22, 24, 25, 28, 29, 31, 33, 34, 37, 39, 44–7, 49, 51–3, 59, 62, 92, 99, 100, 103–8, 116, 118, 119, 121–3, 127, 128, 130, 136, 139, 149, 157, 184
factual, 20, 28, 30, 32, 46, 97, 103, 105, 106, 119, 152

factually, 120
findings, 43, 44, 47–50, 55, 56, 116, 120, 145–7, 150, 161
float, 4, 6, 66–8, 75, 80–83, 89, 92, 95, 108, 132, 158, 170–175
forensic, 80, 89, 99, 103

global, 5, 7, 20, 23–5, 27, 28, 107, 111, 116–17, 119, 144, 145, 149, 158, 162, 173
grant, 22, 135, 157, 160, 161
granted, 4, 17, 84, 112, 154, 156, 173
guidance, 3, 5, 7–9, 99, 104, 106, 107, 112, 113, 139, 153, 165, 166, 181, 187–9

hammock, 89, 173
hypothetical, 116, 118

impact, 1, 2, 27, 32, 36, 38, 39, 50, 57, 59, 75, 78, 103, 104, 108–111, 120, 126, 129, 131, 132, 134, 139, 142, 148–50, 162, 173–5, 183
impacted, 31, 104, 114, 126, 129–31, 134, 136, 142, 145, 188
impacting, 6, 67, 145
imposed, 23, 79, 82, 90, 102, 131, 160, 161, 171, 174, 175
incur, 15, 56, 160, 163
incurred, 4, 5, 7, 16, 21, 22, 26, 31–3, 42, 44, 47–9, 51, 52, 54, 56, 58, 59, 106, 134, 137, 158, 161, 163, 173, 175, 189
incurring, 23, 32, 34, 161
inefficiency(ies), 42, 110, 115, 133–6, 174
inefficient, 41–3, 58, 114, 125, 132, 139, 140, 174, 188
instructed, 21, 33, 40, 47, 86, 87, 154, 156
instruction(s), 7, 14, 16, 21, 22, 33, 97, 102, 107, 111, 113, 118, 119, 130, 141, 144, 147, 153, 155, 156, 158, 159, 163
interrelationships, 172
interrupt, 68
interrupted, 69
interruption, 5, 97, 102, 108, 112, 172, 187
investigation(s), 89, 99, 100, 103–5, 106, 138–40, 147, 150, 152, 172

judge(s), 1, 4, 20, 23, 28, 99, 105, 107, 112, 159, 161
judging, 21
judgment/judgement(s), 9, 11, 13, 18, 20–25, 28, 29, 32–9, 41, 51, 53, 55, 57, 58, 67, 118, 121, 161

lags, 82, 90
late, 19, 20, 27, 29, 35, 36, 38, 41, 43, 46–9, 56, 58, 82, 87, 92, 97, 99, 100, 103, 105, 132, 136, 154, 159, 160, 170, 173, 175, 187
lateness, 36, 38, 187
liquidated, 1, 6, 20, 28–30, 56, 102, 156, 157, 159, 173
loss(es), 1–4, 7, 11–13, 15, 18–20, 24, 25, 27, 33–9, 41, 45, 57, 58, 61, 91, 98–100, 103, 105–9, 112, 113, 115–18, 120–153, 155–8, 161, 162, 165, 166, 169–71, 173, 177, 187, 188

management, 1, 2, 12, 48, 51, 53, 55, 62, 63, 69, 71, 75, 80, 84, 101, 110, 126, 127, 131, 137, 138, 145, 159, 167, 170, 172, 174, 188
manhour(s), 26, 126–8, 133, 139, 143–7, 150, 151
method, 1, 2, 20, 24, 28, 30–32, 37, 45, 49, 62–4, 66, 70, 71, 75, 87, 92, 95, 102, 111, 114–16, 119, 120, 122, 123, 126–8, 134, 136–8, 142, 166, 170, 171, 174, 183
methodology(ies), 12, 18, 99, 104, 106, 107, 110, 135, 136, 162, 172
methods, 1, 2, 5, 20, 24–7, 31, 32, 37, 41, 58, 67, 82, 92, 96, 108, 112, 114–17, 124, 127, 133–52, 162, 187
milestone(s), 45, 75, 82, 88, 171, 174, 177–9
mitigate, 4, 21, 48, 50, 56, 57, 59, 67, 109, 112, 127, 157, 164–7, 174
mitigated, 121
mitigating, 166
mitigation, 3, 4, 6, 11–13, 18, 41, 49, 56, 59, 103, 112, 113, 137, 153–67, 169, 174
modifications, 8, 89, 90, 114
modified, 89, 90, 117, 157, 162
monitoring, 67, 69, 70, 125, 126, 134, 136, 137

network(s), 63–71, 75, 81–4, 90, 91, 169–73, 175

obligation(s), 15–18, 75, 79, 82, 100, 102, 117, 154, 159, 165, 166, 175, 188
obliged, 87, 156, 166, 171, 175
omission(s), 15, 16, 18, 21
omitted, 17, 135
opinion(s), 11, 14, 16, 17, 35, 82, 104, 106, 139, 141, 142, 145, 147–152
output(s), 26, 84, 86, 95, 115, 123, 125, 126, 130, 142, 145, 164
overlap, 102, 163
overlapping, 65, 143, 173

overrun(s), 2, 29, 100, 137, 138, 145
overtime, 34, 51, 116, 129, 130, 139, 143, 149, 158, 161, 162

payment(s), 2, 5, 15, 19, 20, 22, 23, 25, 26, 30, 34, 112, 117, 118, 128, 134, 153–6, 170, 187
planned, 4, 6, 34, 41–3, 45, 46, 48, 58, 61, 62, 67–70, 75, 77, 79, 81, 87, 89, 92, 104, 105, 112, 119, 122, 123, 125, 126, 129, 131, 136, 137, 142, 144, 145, 149, 153, 154, 156, 158, 163, 165, 166, 172, 174, 183, 188
planning, 8, 21, 41, 43, 58, 61–98, 100, 101, 103, 114, 130, 132, 137, 167, 169, 175, 177
predecessor(s), 83, 89, 90, 171–3
prevent(s), 1, 2, 14, 47, 59, 90, 117, 165–7, 188
prevented, 6, 49, 167
prevention, 6, 15, 16
preventive, 167
principle(s), 3–7, 17, 22, 24, 47, 52, 69, 100, 112, 113, 116, 154, 166
productive, 31, 48, 56, 125, 129
productively, 111
productivity, 1–4, 11–13, 18, 26, 34–8, 43, 58, 61, 66, 98, 99, 101, 107–10, 114–17, 119, 120, 123–153, 158, 161, 162, 169, 172, 177, 187, 188
program(s), 2, 24, 62–5, 90
programme(s), 2–4, 7, 8, 25, 26, 31, 32, 36, 38, 40–51, 56, 57, 59, 61–98, 104, 112–114, 119–22, 132, 137, 140, 143–5, 149, 155, 163, 166, 167, 169, 170, 172–5, 177–185, 188
programmed, 26, 62, 172
programming, 8, 34, 45, 61–78, 97, 103, 104, 132, 163, 174, 182, 183
progress(es), 2–4, 7, 8, 11, 14, 15, 17, 26, 27, 34, 36, 39, 42, 43, 47, 61, 66–70, 78, 83–5, 87–97, 103–105, 107, 108, 112–14, 126–8, 134, 137, 138, 146, 150, 151, 158, 160, 161, 163–5, 169–75, 178, 180, 181, 187
progressed, 35, 104, 146
prolongation, 1, 4, 5, 7, 28, 36, 55, 99, 100, 105, 106, 113, 157, 175
prolonged, 130, 158
proof, 24, 25, 28, 31, 107, 116, 118, 136
Protocol, 1, 3–9, 99, 104–7, 112–14, 153, 154, 165, 174, 175, 187–9

quantification, 2, 18, 28, 41, 58, 109, 114–17, 135
quantified, 27, 111, 137, 174
quantify, 1, 2, 108–10, 117, 127, 133, 136, 159
quantitative, 138, 146
quantum, 31, 42, 136, 158

reality, 2, 24, 28, 30, 42, 47, 55, 118, 120
reason, 30, 41, 43, 44, 46, 50, 52, 57, 58, 81, 85–7, 90, 116, 119, 120, 134, 155, 156, 158, 161, 188
reasonable, 3, 4, 6, 11, 14, 15, 17, 18, 21, 25, 31, 32, 47, 48, 50–52, 54, 56, 57, 59, 67, 83, 102, 112–16, 134–6, 139, 156, 158, 160, 161, 163, 165–7, 189
reasonableness, 83, 100
reasonably, 11, 14–18, 46, 109, 134, 135, 156, 160–163, 165
records, 2–5, 7, 25, 36, 41, 43, 58, 61, 69, 88, 91–8, 105–7, 109–13, 115, 116, 126, 134, 139, 141, 144, 145, 147, 148, 150, 162, 163, 188, 189
recovery, 22, 25, 30, 31, 42, 46–51, 56, 57, 59, 108, 109, 122, 123, 138, 139, 142, 157–61, 166, 170
reimbursable, 7, 123, 125
reimbursed, 15
reimbursement, 152
relationship(s), 1, 33, 62, 65, 66, 68, 70, 71, 75, 82, 83, 90, 100, 129, 158, 175
relevant, 3, 9, 11, 14–17, 21, 23, 33, 39, 53, 75, 85, 103, 105, 115, 116, 157, 165, 166, 188
reprogramme, 21
reprogramming, 21, 67
reschedule, 21
rescheduling, 21
resource(s), 1, 4, 21, 27, 30, 41–3, 48–53, 55, 57, 58, 65, 67, 80, 81, 98, 111, 112, 117, 120, 129, 131, 132, 142, 156–8, 163–6, 169, 170, 172, 174, 175, 180, 188
resourcing, 121
restraint(s), 66, 71, 82
retrospective, 5, 99, 104–6, 120
retrospectively, 107, 117
revised, 20, 40, 85–92, 145, 151, 155, 158
revising, 69, 80, 85, 90
revision(s), 4, 21, 40, 61, 79, 84–91
rework, 110, 129, 130, 132, 139–43, 148–50, 158
risk(s), 4, 5, 34, 80–82, 91, 101, 102, 105, 107, 112, 115–17, 165–7, 170–175

schedule(s), 3, 14, 16, 28, 30–32, 35, 41, 43, 58, 62, 77–9, 84, 91, 100, 126, 127, 132, 137, 138, 147, 154, 155, 159, 161, 170, 175, 177–85
schedulers, 78
scheduling, 66, 80, 177, 182, 183
slippages, 138
software, 2, 8, 24, 63–9, 75, 78, 89–91, 103, 114, 170, 174, 175, 181
subjective, 2, 104, 121
successor(s), 65, 66, 81, 83, 90, 172, 173
suspension(s), 8, 15, 16, 102, 114, 160

Index 195

time(s), 1–9, 11, 14, 15, 17, 18, 20–22, 24, 28–30, 32, 36, 38, 40–43, 45–48, 50–58, 61–3, 65–7, 69–71, 74, 75, 77–88, 90, 92, 95–108, 112, 115, 118, 119, 121, 122, 124–7, 129–35, 137, 140–142, 144, 148, 149, 151–64, 166, 167, 169–75, 177, 180
timely, 69, 101, 132, 160

unproductive, 33, 37, 131, 139–41, 144, 148–151
unproductively, 20
update(s), 8, 61, 68, 69, 82–4, 87–92, 114, 146
updated, 4, 8, 35, 69, 87, 88, 112, 114, 132, 175
updating, 8, 87–9, 91, 114

validation, 110
valuation, 4, 7, 19, 27, 30, 33, 36, 170, 189
value, 31, 37, 41, 42, 58, 86, 92, 100, 115, 116, 124, 126–8, 133, 137–52, 156
variation(s), 2, 4, 5, 7, 14, 16, 19, 20, 22, 23, 26–30, 33, 36, 42, 47, 49, 63, 97, 105–7, 110, 115, 118, 141, 144, 147, 155, 170, 175, 189

varied, 25
verification, 20
verified, 43

weather, 15, 46, 96, 98, 118, 129, 139, 142, 144, 148, 150, 164, 188
workflow, 83, 171
workforce, 187
working, 4, 5, 7, 19, 20, 28–30, 33–35, 37, 39, 40, 42–51, 56, 57, 70, 75, 82, 87, 92, 95, 106–8, 111–14, 116, 130, 131, 134, 136, 139, 140, 142, 143, 145, 149, 158, 159, 161–7, 174, 183, 187, 188
works, 2–4, 8, 9, 11, 14, 15, 17, 19, 33–6, 38, 40, 41, 45, 48, 56, 58, 62, 65–8, 79–82, 84–87, 89, 90, 92, 99, 103–105, 107, 109, 112–14, 116, 119, 141, 153–5, 157, 158, 163–7, 169–75, 180, 181
workweek, 127, 134

Also Available from Wiley Blackwell

Evaluating Contract Claims
Second Edition
Peter Davison & John Mullen
978 1 4051 5920 3

Delay Analysis in Construction Contracts
PJ Keane and AF Caletka
978 1 4051 5654 7

Building Contract Claims
Fifth Edition
David Chappell
978 0 470 65738 6

Building Contract Casebook
Fifth Edition
Michael Furmston
978 0 470 65592 4

The JCT Standard Building Contract 2011
David Chappell
9781118819753

A Practical Guide to the NEC3 Engineering and Construction Contract
Michael Rowlinson
978 1 4443 3688 7

The NEC3 Engineering and Construction Contract: A commentary
Second Edition
Brian Eggleston
978 0 632 05386 5

The FIDIC Form of Contract
Third Edition
Nael Bunni
978 1 4051 2031 9

Building Law Encyclopaedia
David Chappell, Michael Cowlin and Michael Dunn
978 1 4051 8724 4